Timber Building in Britain

Timber Building in Britain

R. W. Brunskill

London
Victor Gollancz Ltd
in association with
Peter Crawley
1985

R. W. Brunskill

TRADITIONAL BUILDINGS OF BRITAIN:
An Introduction to Vernacular Architecture

TRADITIONAL FARM BUILDINGS OF BRITAIN

ILLUSTRATED HANDBOOK OF VERNACULAR ARCHITECTURE

VERNACULAR ARCHITECTURE OF THE LAKE COUNTIES

ENGLISH BRICKWORK (*with Alec Clifton-Taylor*)

Except where advertised as open to the public all the buildings mentioned or illustrated in this book are in private occupation and readers are asked to respect that privacy.

First published in Great Britain 1985
by Victor Gollancz Ltd,
14 Henrietta Street, London WC2E 8QJ

British Library Cataloguing in Publication Data
Brunskill, R. W.
  Timber building in Britain
  1. Architecture—Great Britain—History
  2. Building, Wooden—Great Britain—History
  I. Title
  721'.0448   NA4110

ISBN 0–575–03379–7

Designed by Harold Bartram
Filmset and printed in Great Britain by
BAS Printers Limited, Over Wallop, Hampshire

Contents

List of photographs

39 Braces, house at Sutton Courtenay, Berks. (PSC)
40 Braces, building in West Stockwell Street, Colchester, Essex (MUSA)
41 Braces, Capons Farm, Cowfold, Sussex (PSC)
42 Windbraces, solar roof, Fiddleford Mill, Dorset (PSC)
43 Passing brace, Grange Farm Barn, Coggeshall, Essex (RCHM)
44 Flying bressummer, Monks Barn, Newport, Essex (PSC)
45 Braces, Old School House, Bunbury, Cheshire (PSC)
46 Bracket, Trewern, Montgoms. (RCHMW)
47 Carpenters' marks, gatehouse at Bolton Percy, Yorks. W.R. (RCHM)
48 Carpenters' marks, Melville House, Fife, Scotland (RCAMS)
49 Mathematical tile cladding, building in West Street, Faversham, Kent (PSC)
50 Pargetting, house in Fore St., Hertford, Herts. (PSC)
51 Weather-board cladding, house at Smarden, Kent (PSC)
52 Slate cladding, The Nunnery, Dunster, Somerset (PSC)
53 Cove, Bents Cottage, Mottram St. Andrew, Cheshire (MUSA)
54 Crown post, Durham House, Great Barfield, Essex (RCHM)
55 Cruck truss, dairy farm, Mobberley, Cheshire (MUSA)
56 End cruck (hip cruck), barn at Corrimony, Inverness-shire, Scotland
    (RCAMS)
57 Jointed crucks, barn at Preston Plucknett, Somerset (RCHM)
58 Raised crucks, barn at Barton Farm, Bradford on Avon, Wilts. (PSC)
59 Two-tier crucks, barn at Barton Farm, Bradford on Avon, Wilts. (PSC)
60 Truncated crucks, cottage at Styal, Cheshire (MUSA)
61 Cusping, Old Vicarage, Glasbury, Radnors., Wales (RCAHMW)
62 Fastenings, Herbert Warehouse, Gloucester (RCHM)
63 Interrupted sill, house at Thornhill Lees, Yorks. W.R. (RCHM)
64 Hammer beam, Church of St. Wendreda, March, Cambs, (PSC)
65 Interrupted tie-beam, Banister Hall, Walton-le-Dale, Lancs. (RCHM)
66 Jettying, St. John's Alley, Devizes, Wilts. (PSC)
67 Jettying, cottages at Cerne Abbas, Dorset (PSC)
68 Jettying, Myddleton Place, Saffron Walden, Essex (PSC)
69 Jointing, barn at Grange Farm, Coggeshall, Essex (RCHM)
70 Scarf joint, North Warehouse, Gloucester (RCHM)
71 King post roof truss, former theatre, Dorchester, Dorset (RCHM)
72 King post roof truss, Howard St. Warehouse, Shrewsbury, Salop. (RCHM)
73 Boarded infill, Coppice Farm, Ashley, Cheshire (MUSA)
74 Brick nogging, Old Manor House, Romsey, Hants. (PSC)
75 Square panelling, Pembridge, Herefords. (PSC)
76 Quatrefoil panelling, Lydiate Hall, Lydiate, Lancs. (RCHM)
77 Muntin and plank panelling, Berllan-deg, Llanhenog, Mon. (RCAHMW)
78 Pegs, barn at Corrimony, Inverness-shire, Scotland (RCAMS)
79 Angle post and jettying, The Chantry, Sudbury, Suffolk (PSC)
80 Purlins, barn at Avebury, Wilts. (PSC)
81 Clasped purlins, house at Carlton Husthwaite, Yorks. N.R. (RCHM)
82 Princess posts, Howard St. Warehouse, Shrewsbury, Salop. (RCHM)
83 Queen post roof truss, Albert Warehouse, Gloucester (RCHM)
84 Crown post rafter roof, Church of St. Michael & All Angels, Codford, Essex
    (RCHM)

Key to credits

PSC	Peter Crawley
MUSA	Manchester University School of Architecture Collection
JMcC	John McCann
RCAHMW	Royal Commission on the Ancient and Historical Monuments of Wales
RCAMS	Royal Commission on Ancient and Historical Monuments of Scotland
RCHM	Royal Commission on the Historical Monuments of England
WJS	W. J. Smith

In this list and throughout the book the traditional county names and locations have been used.

Preface and acknowledgements

This book, like so much of my work, stems ultimately from the inspiration of Professor R. A. Cordingley. In 1961 he published in the *Transactions of the Ancient Monuments Society* an article entitled 'British Historical Roof-types and their Members'. The article appeared not long before his death in 1962, but Cordingley had already begun to collect photographs of church roofs with the intention of producing an expanded and much more comprehensive work on the subject. I had played some small part in the preparation and illustration of Cordingley's article and always hoped it would be possible to take the subject further especially in the matter of the glossary. Although the very considerable amount of study and publication which has taken place in the past twenty years or so has called into question the basis of Cordingley's main argument in roof classification, nevertheless the collection of terms in a glossary and the illustration of roof-types and other aspects of timber-frame construction in line diagrams continues to be useful and there has been a steady demand for offprints of Cordingley's article.

It seemed to me that there was a need for an up-to-date summary of current understanding of timber construction as a whole which would be based on an illustrated glossary of techniques and terms. Through diagrams and illustrations, as well as text, this would make the material scattered in highly technical books and articles more readily available in summary form to the rapidly increasing band of enthusiasts for all forms of traditional building, especially those whose interest had been aroused by the one or two excellent introductory booklets now available.

This book also owes a debt to the inspiration of Alec Clifton-Taylor. His work on building materials is well-known and has been very influential. I have been privileged to collaborate with him in expanding some of the material in *The Pattern of English Building* when we took one of the main constructional materials as the subject for *English Brickwork*. The pre-eminent building material, stone, has more recently been the subject of a book in which Alec Clifton-Taylor collaborated with A. Ireson, the well-known expert on stone, quarries and the mason's craft. I would like to think that a book on the tradition of timber building in this country would complete the expansion of the three main materials covered in Alec Clifton-Taylor's major work.

However, it is only possible to consider the preparation of a summary volume because of the existence of the long, and recently much-increased, study of timber-framed structures. A book such as this must lean heavily on the published work of experts and especially of C. A. Hewett, J. T. Smith and F. W. B. Charles. Hewett's vast and careful researches in Essex and elsewhere have transformed our knowledge of the theory and techniques of timber building and especially roof construction. The wide-ranging studies published by J. T. Smith over the past thirty years have placed both general study and detailed investigation in context. F. W. B. Charles has brought experience of conservation and repair to intimate study especially of cruck-framed buildings. Among the other students of vernacular architecture whose published work has been indispensible in producing this book, pre-eminent are N. W. Alcock, M. W. Barley, E. A. Gee, R. Harris, S. R. Jones, J. McCann, E. Mercer, and P. Smith. In addition, many more members of the Vernacular Architecture Group have been as instructive as ever.

Like Cordingley, several authors included glossaries in their works and these have all been useful. However, from Moxon in the seventeenth century, through books such as those by Neve and Nicholson in the eighteenth century and text-books such as that of Tredgold in the nineteenth century and culminating in the great *Dictionary* of the Architectural Publication Society has come much of the material which has formed the basis of the glossary in this book.

Most of the photographs have been specially taken for the book by Peter Crawley and I am again grateful for his industry, enthusiasm and skill. Others are from the National Monuments Records of England, Scotland and Wales and I am happy to acknowledge the great help I have obtained from the staff of these three collections. The diagrams I have prepared myself.

R. W. Brunskill
Wilmslow, December 1984

Introduction

This book is about carpentry, the craft which has been variously described as 'the theory and practice of framing timber' and 'the art of employing timbers in the construction of buildings'. Traditionally, carpentry involves timber and timber is not quite the same thing as wood. By timber is meant the material which was used for the construction and repair of the structural parts of a building, a bridge or a ship. In the case of a building this means such items as the roof members, the posts and beams of the wall and the joists and boards of the floor. Wood, on the other hand is the material used for furniture or tools or burnt as fuel on a hearth or converted into charcoal for a barbecue. The distinction between timber and wood was a very important one, enshrined in manorial customs and carefully set out in the lease which formed a contract between a landowner and a tenant farmer. Buildings, therefore, were constructed of timber though they may have been finished with woodwork.

Building in timber was the province of the carpenter though the joiner might be involved in the fitting-out of a timber structure and, in time, the cabinet-maker developed his craft in finishing it. The craft of joinery developed from that of upholstery but eventually the joiner took over the finer parts of working in wood. Thus screens and panels were joiner's work and as the production of doors and windows became more complicated, requiring specialist tools and a certain skill, they were made by the joiner rather than the carpenter. Fixed furniture, such as cupboards, might be made by the joiner but the cabinet-maker emerged as the specialist in loose furniture, often of a very delicate construction. But at all times structural timber was worked by the carpenter.

Most of the large-scale work of the carpenter has been set aside, though even now a few large timber bridges survive in this country—such as the one which carries the railway over the Mawddach Estuary at Barmouth in North Wales. There were once many more to demonstrate the sweep of the carpenter's vision but although they have their own interest, as works of engineering rather than building they have not been included here.

Examples of timber building have been taken from various parts of Great Britain though inevitably nearly all are from England. There are many timber-framed buildings in Wales but they seem to share common traditions with adjacent parts of England. Scotland has fewer surviving buildings based on timber construction and these seem mainly to use variations on techniques which may also be seen in England. In both countries, however, there are alternative terms for some of the items appearing in the glossary.

Although some of the highest achievements of the artists in timber have been considered and illustrated, the book concentrates on the more humble and everyday uses of timber. Great works such as the roof of Westminster Hall deserve their own separate study. It is in the everyday buildings, houses and parish churches, market halls and barns that the work of the carpenter can be most readily observed and appreciated. Of these buildings the secular predominate over the ecclesiastical. One has to turn to churches for some of the early roofs and the parish churches of East Anglia, for instance, provide some of the most spectacular pieces of carpentry, but completely timber-framed churches are few whereas timber-framed houses are many and may be seen in most parts of the country. Even when surrounded by stone walls in the Lake District, the Pennines or the Brecon Beacons one ought to be conscious of the timber work so often concealed by these materials.

Timber has been used for building in this country from the earliest times right up to the present day. Some mention will be made of such use of timber as can only be detected through excavation. There will be a note on recent developments in carpentry. But the bulk of the material on which this book is based comes from the late Middle Ages through to the nineteenth century, about seven or eight hundred years and of this probably the sixteenth, seventeenth and eighteenth centuries are the most interesting. Medieval carpentry generated a flare of ingenuity in the sixteenth and seventeenth centuries; some of the most ambitious work was done in the eighteenth and early nineteenth centuries just before iron and steel took over from timber in many roles. However, some work remained for the carpenter and in our own times there has been a resurgence in the use of timber, of different varieties and employing different constructional techniques replacing a moribund but not quite forgotten carpentry tradition.

We are concerned in this book with mainstream traditional carpentry. Individual designers in timber have been known from the medieval period onwards. Nowadays structural timber whether assembled in a factory or put together on the building site has to be designed by engineers and checked by building inspectors; the old rules of thumb are no longer acceptable. But fortunately, much survives of traditional carpentry and it is on this work rather than on the individual achievements of a professional timber engineer, of whatever century, that this book is based.

Interest in timber building lies both in the elements themselves and the ways in which these elements were brought together to produce walls, roofs, floors or complete buildings. The book therefore begins with a section, mainly of text, covering walls, cruck construction, roofing for narrow and wide spans, floor construction and deals in turn with predecessors to our surviving structures, and with more recent developments.

This is followed by the main part of the book which consists of a glossary in which technical terms are described and illustrated by sketches and photographs. The glossary attempts to bring together terms which are traditional and those which have been devised by recent scholars to deal with items or arrangements for which no traditional terms are known. Often there are several variations even among traditional forms: variations differing from region to region, and variations as terms became archaic and were superseded by others. To some extent there has been some duplication, several terms having been used for the same item, some words meaning different things to different writers. The glossary, therefore, does include a measure of selection. A sensible choice has been intended, but in a craft rather than a science choice has tended to be by use rather than logic.

The third part draws elements together again with the aid of photographs of buildings and parts of buildings arranged generally in chronological sequence. While the first part of the book explains the operations of the carpenter's craft and the second deals in detail with the elements with which he works, the third part attempts to return to the complete work of the carpenter. The approach is roughly chronological though many buildings illustrate work of several periods and some techniques remain static for a long time while others are changing more rapidly. The third part also attempts to concentrate on the buildings and parts of buildings which are readily visible rather than those parts, often of great technical interest, which are usually hidden and whose significance has already been explained.

Timber building has been studied recently by archaeologists, architects and historians, professional and amateur. Each has brought his own techniques of recording, analysis, interpretation and explanation. For each one the surviving buildings and their documentary trail can help to illuminate his own subject. Academically, study of surviving carpentry can add so much to the body of knowledge about building. Practically, the careful study of timber buildings is essential for those specialists charged with their preservation or restoration. However, the survival of timber buildings depends ultimately on the appreciation of owners, tenants and members of the public and appreciation in its turn depends on the understanding which it is hoped this book will help.

Part One:
The Construction of Timber Buildings and Roofs

1. The Old Shop, Bignor, Sussex

This timber-framed house is of the 'Wealden' type. It has three structural bays with the upper parts of the two outermost bays jettied forward. The studs and rails of the walls make panels of various shapes, many of which are now infilled with nogging of brick or flint.

Braces rise from the sills to meet the wall-posts. A flying bressummer, braced from upper wall-posts, carries the eaves between the jettied upper storeys. The hipped roof probably conceals crown post and collar purlin roof construction.

Introduction

As a general introduction to the subject, Part One begins with some common-sense observations on structural theory: the forces acting on timber structures and the ways of counteracting those forces. Then there are some general observations on the predecessors to our surviving timber buildings, the special nature of their constructional problems, the limited means available to deal with those problems and the influence the various solutions had on later examples. Evidence is largely from excavation, and new discoveries and new interpretations inevitably affect our attitudes to this archaic work. Next the problem of providing the building material is considered: how trees were selected, how they were cut down, how they were converted into useful members, how they were transported to the building site and how they were made available for the carpenter's work. This is followed by a brief examination of the range of jointing techniques—the means whereby the carpenter was able to assemble his separate pieces of timber so as to put together a structure which would serve the intended purpose.

Part One continues with a summary of the problems and solutions associated with cruck construction, that simple and sometimes crude assembly of timbers which tackled the problem of roofing and walling simultaneously. Next the more common box-frame and post-and-truss assemblies are considered whereby walls of primary and secondary timbers were erected and roofs raised over them so as to have considerable or slight relationship between the box of the wall and the lid of the roof. The special problems of wide-span roofs are briefly explored together with the standard solutions involving intermediate supports or complicated roof trusses utilising built-up timber members and eventually the introduction of wrought iron and cast iron to supplement the strengths of the available timber.

The third section of Part One covers the less specialist uses of timber in partitions and flooring. Mention is also made of timber decoration; that which comes from the arrangement of timber members as in forming decorative panels in walls or foliated designs in roofs; that which comes from the addition of decorative work such as barge boards or finials and that which comes from the moulding of the timbers themselves.

Inevitably this introductory material requires the use of many technical terms. They are defined, described, discussed and illustrated in the glossary which follows to form Part Two.

Structural theory

Traditional carpentry was pragmatic; the carpenters dealt with each problem as it arose. Modern scientific carpentry is based on theory; problems are anticipated and designs are prepared to deal with each problem before it arises. Nevertheless, traditional and scientific carpentry are related to each other; pragmatic solutions might precede the scientific explanation of why a solution actually works, structural theory is to some extent founded on the observed effect of using timber in traditional ways. There has been a long overlap between the gradual abandonment of traditional carpentry in the middle of the twentieth century and the gradual introduction of scientific carpentry, possibly from the mid-seventeenth century and certainly from a late eighteenth-century date. In any case, potential structural problems existed whether solutions were traditional or scientific.

Timbers put together to form buildings or parts of buildings have to deal with three main types of loading: superimposed loads, self-weight and wind loads. Superimposed loads are those resulting from the functions performed by the timber-framed building; they include the weight of people, furniture, crops, equipment in a structure as well as the weight of thatch, slate, tiles or other roofing materials, and panels or cladding which keep the weather out of the walls. Self-weight represents the loads imposed by the timber itself: that weight of floor-joist, for instance, which maintains its integrity before anyone imposes a load on the floorboards. Wind load consists of the forces imposed on the structure from outside; it may result from the sheer pressure of a high wind on the weather side of a building or the suction affecting the lee side as a result of that wind. Effects similar to wind load may result from other factors such as the movement of loose material in a barn or warehouse. (d1a,b)

These three sorts of loading affect timber buildings in three main ways: through bending, through shear and through structural movement.

Through bending a member is likely to collapse between supports. Superimposed loads, self-weight, or a combination of these, result in bending, sagging or deflection of the member and in extreme cases may result in collapse because the bending tendency is greater than the fibres of the timber can bear. The most obvious and familiar type of bending is associated with horizontal members whereby a joist or tie-beam may sag and collapse when over-loaded or (when subject to too much decay) may become inadequate in size for its intended load. Members may also fail through sideways or lengthways bending, however, in the manner one usually associates with buckling. A beam such as a wall-plate may be overloaded in such a way that it is adequate against sagging but fails through collapsing sideways. A post may be of sufficient dimensions to deal with a crushing load applied vertically but fail because that load causes the post to buckle and so collapse. (d2a,b,c)

Through shear a member is likely to collapse at its supports. A beam may

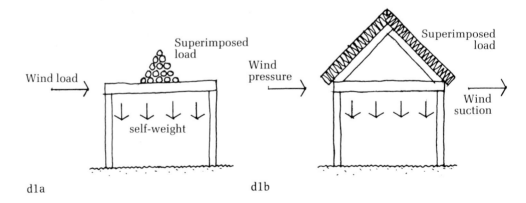

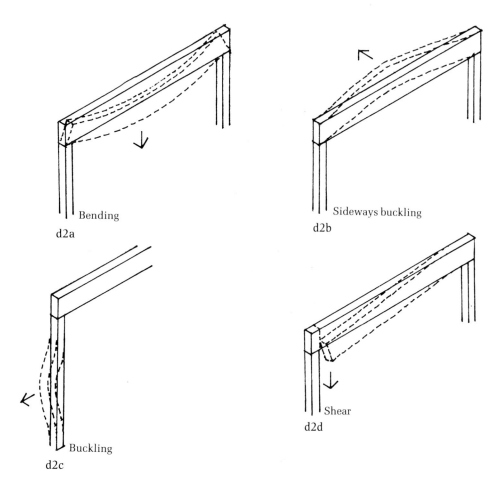

Bending
d2a

Sideways buckling
d2b

Buckling
d2c

Shear
d2d

be deep enough at its centre to avoid excessive deflection but still fail because it is too slender at its supports, either because it was inadequate in the first place or because vital timber was cut away in making a joint, or simply through decay. (d2d)

Structural movement may be lengthways, sideways or involve the corkscrew effect of both at once. A common type of such movement is called racking and involves the movement longitudinally of a series of members in a wall or roof. Racking may be inconvenient as when doors do not fit or windows do not open, but in extreme cases it may result in the collapse of the building. Sideways movement as if a building were leaning drunkenly may result from collapse of foundations, loss of strength in vital members or through wind loads. The twisting or corkscrew movement of a building—a combination of racking and sideways movement—imposes unexpected loads in many parts of a structure and may lead to general weakness or collapse. (d3a,b,c,d)

Sometimes these forces act separately but usually they act together and in any case problems of one sort of loading affect a structure in other ways. Thus an overloaded beam may deflect so much as to cause failure in shear, which then unbalances the structure which eventually collapses through racking.

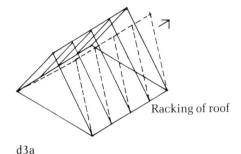

Racking of roof

d3a

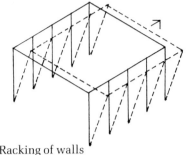

Racking of walls
d3b

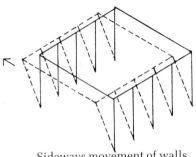

Sideways movement of walls
d3c

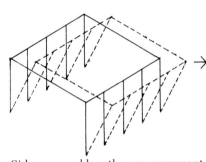

Sideways and lengthways movement
(corkscrew)
d3d

The skill of the carpenter involved the use of his material in such a way as to counteract collapse through bending, shear or movement. He could use the shape of his building, the selected size of members, the type of supporting or propping members, triangulation and appropriate jointing to produce a stable and long-lasting structure.

One way of shaping a building to resist wind forces was to add extensions or aisles at the sides or end. In a fully-aisled barn, for instance, the aisles at both sides and both ends acted as props to help keep the central structure (which might be quite slender) in a stable condition. A hipped roof could reduce the chances of failure of the roof members through racking: any tendency for the roof to collapse lengthways was counteracted by the hipped ends leaning inwards. A building with projecting wings, H-shaped or U-shaped or even T-shaped could be stabilised as the short wings running in one direction prevented collapse of the main body running in the opposite direction which in turn gave stability to the wings. (d4a,b,c)

Although timber is quite strong in compression and straightforward tension it is vulnerable to that tension which results from bending, twisting or shear and so splits the fibres. It appears that, traditionally, carpenters attempted to keep members in compression as much as possible. One way was to use them as props. Thus if a beam such as a tie-beam seemed likely to sag and so collapse through tension tearing the fibres of the bottom of the member it would be propped. Such props might be vertical, as a post, or inclined, as a brace or strut. Wherever possible the props were kept short so that they would not themselves fail through buckling, or they were given extra horizontal or diagonal propping to avoid such failure. Struts or props might act in different planes: a roof could have its purlins propped by wind braces, its principal rafters propped by diagonal braces or its wall-plates propped by knee braces. At the same time a wall-plate spanning between posts might be considered as propped at intervals by wall studs while these and the posts were saved from buckling by means of the rails. Traditional carpentry may be considered, to a large extent, as a system of bending and propping.

However, consciously or unconsciously, traditional carpenters also made use of triangulation. Any tendency for a structure to collapse under wind loading could be countered by the use of diagonal members to make up triangles. Thus the very braces which acted as struts in reducing the unsupported length of a wall-plate, tie-beam, principal rafter or purlin triangulated the members one to another. (d4d)

Nearly all timber buildings are based on squares or rectangles or combinations of the two. A timber frame, square on plan, braced at the corners is likely to be a strong piece of construction, well able to cope with superimposed loads, self-weight and wind loads. An elongated rectangle, however, is likely to fail through the outward collapse of its long walls unless the tops of those walls are tied together in some way. This is usually done by tie-beams deep

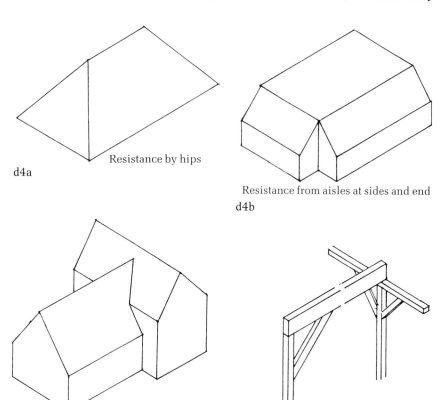

Resistance by hips
d4a

Resistance from aisles at sides and end
d4b

Resistance from wing
d4c

Resistance by triangulation
d4d

enough to support their self-weight and any superimposed load. Such tie-beams divide the rectangular building into bays along its length, a particularly convenient division if intermediate posts are placed between corner posts and underneath each tie-beam. Such tie-beams may also serve as the basis for transverse roofing members or parts of trusses so that both roof and walls are divided into similar structural bays. In cruck construction where posts and roof principals are one this bay division occurs automatically. Thus a rectangular building may have walls of various members held together by tie-beams, or walls divided into structural bays filled in by non-structural studs and rails and corresponding to roof bays, or have cruck trusses at intervals forming bays.

Pre-framing

Before considering the surviving timber-framed structures we should ponder briefly on their predecessors, known only from excavation, legend, description or partial survival. These are the structures which relate to an abundance of timber rather than a scarcity, to the periods and places in which timber was a weed to be eliminated or an obstacle to the cultivation of land, to circumstances in which prodigal rather than economical use of timber was to be expected.

As settlers gradually penetrated these islands they had eventually to move away from those limited tracts in which a certain elevation and a certain soil condition produced a thin sparse tree cover into those much more extensive areas in which thick forest covered the land. Pollen analysis has shown that most of this country was once thickly forested: areas now bare, such as the mountains of England, Wales and Scotland, were naturally forested before clearance removed timber and woodland alike and animal pasture prevented regeneration. At the same time the ill-drained but fertile clay lowlands now producing our best pasture were formerly thickly forested. Before any of this land could be farmed it had to be cleared: trees chopped down, stumps grubbed up, undergrowth eliminated. Any use for the timber which helped remove it from the prospective fields would ease cultivation. Building was such a use.

Somewhat similar conditions were faced by settlers in lands which had previously been uninhabited or occupied only by hunters. Contemporary accounts of the settlement of wooded North America from the seventeenth century through to the twentieth show that even well-equipped settlers in modern times had to girdle trees until they died and could be pulled up, or chopped down, and stumps grubbed up with the aid of oxen or blown up by means of gunpowder. The settler's log cabin was a means of using up timber as well as providing shelter.

In circumstances such as these timber logs were used to produce walls of mass construction, the logs serving rather as if they were elongated bricks or stones whether placed horizontally or vertically. In the absence of good quality tools, the more solid the wall and the fewer the joints, the easier could the building be erected. Horizontal logs could be piled one on top of another

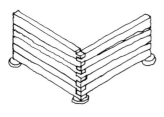

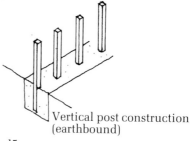

Horizontal log construction
d5a

Vertical log construction
(earthbound)
d5b

Vertical post construction
(earthbound)
d5c

after the fashion of the log cabins of Scandinavia, Alpine Europe and North America, with simple corner joints and doors and window openings cut out of the solid walls. Needing no foundations, such walls would leave little or no archaeological evidence but documentary evidence shows that in special circumstances something like this technique survived into the sixteenth century in Northern England. (d5a) However, unlike the softwood timber of the countries associated with horizontal log construction, the hardwood timber of Britain would be unlikely to produce long, uniformly and only slightly tapering tree trunks even in primeval forests. Curved and bent timber and short lengths of straight timber were more appropriate to vertical log construction. In this technique, storey-height walls could be obtained by using rows of logs placed upright in a trench and extending to the limited height which was all that was required. The trench, when back-filled, gave stability to the logs and so helped to provide a satisfactory wall. However, such a wall was not long-lasting. Timber deep in the trench or timber high in the wall was reasonably safe from decay but the base of the wall as it rose from the ground was very susceptible to the alternate wetting and drying which causes decay and might begin to fail after only a few years. The disturbed earth of the trench and the decayed vegetable matter of the lower part of the logs provide good archaeological evidence for this type of construction. (d5b,c)

In either form of log construction the wall was well-designed to counter crushing under vertical loads, sideways buckling from too thin and unrestrained a length of wall, cross-ways or lengthways collapse through racking. A stable wall could easily be provided. The form of roof construction is largely unknown but coupled rafters producing light loads at close intervals seem most appropriate and not unlikely.

However, the period of clearance was limited and once logs had to be hauled from the forest edge to a site on cleared land economy in use of timber became increasingly important. Horizontal log construction gave little chance to save timber but vertical log construction allowed for the timbers to be placed further and further apart, the infill changing from 'chinking' clay or mud in narrow gaps between adjacent logs to panels or walls of such material as fewer logs entailed wider gaps. The logs could still be given stability, rising from holes or back-filled trenches like so many large fencing posts and propped with temporary or permanent inclined stays as the earth was back-filled and consolidated in the post-holes.

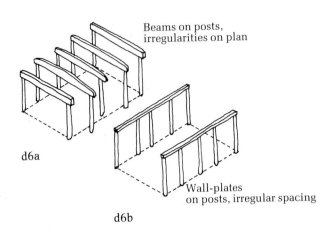

Beams on posts,
irregularities on plan

d6a

Wall-plates
on posts, irregular spacing

d6b

In such construction, roofing problems began to occur. There was no longer a continuous upper surface on which the feet of rafters could bear, instead it was necessary to introduce a continuous wall-plate, linking the tops of all the posts, or to allow for linking opposite posts with some sort of tie above which a roof could rise. In either case the question of the alignment of the tops of the posts became important. Many excavations can show evidence of post-holes in a line but few can demonstrate the use of a straight line and the question of how the upper points of these posts were linked and roofed is still far from satisfactorily answered. One theory is that posts were arranged to suit tie-beams, fairly closely spaced sets of two posts and a tie-beam being linked together longitudinally by light wall-plates or by the longitudinal members of the roof construction. Such a theory would suggest that what we now call reversed assembly—the placing of tie-beam below wall plate—is a more ancient technique than normal assembly. (d6a,b)

Between the posts, whether closely or widely spaced, was the non-structural wall and excavation suggests that almost any material able to support its own weight could be introduced. Thus there were walls of rough masonry, of clay, of wattle and daub, of large storey-height panels and eventually of the small panels with which we are now familiar.

Once it was realised that vertical timbers need not be earth-fast to ensure stability but could be made stable through framing, and once the tools and jointing techniques essential for framing had became available, the type of timber-braced construction which shows the carpenter's craft could be produced.

Operations

The carpenter needed a material which was readily available, easily and cheaply selected or converted to dimensions adequate and manageable, easily worked but durable and weather-resistant when in use. The woodlands and hedgerows could be made to provide suitable timber, especially oak. The carpenter had to compete with other uses for some parts of the timber but other parts he could share. The selection, conversion and provision of structural timber was as much a part of the traditional craft of the carpenter as the choice of meat at the market is part of the skill of the modern chef.

For centuries after the primeval forests had been largely cleared and their timber used or squandered the cultivation of timber and wood as a crop was maintained, a practice only discarded when other materials or cheap imports became available, and not entirely even then. Timber was used for building, wood was used for fuel, if of poor quality, or for joiner's work if of superior quality, brushwood and the waste from timber was used for hurdles and similar short-lived artefacts. A wood or coppice could be cultivated by a medieval community to produce a regular supply of thin coppice wood for fires or charcoal-making and tall standards for timber building. Each individual tree could itself be made to produce bark for tanning, branches for burning as well as timber for building. It has been estimated that the woods and hedgerows of West Suffolk, properly managed, cut and replanted according to a regular cycle, could produce, during the medieval period enough structural timber for two houses to be built in each parish every five years as well as adequate wood for other operations and uses. Following the Dissolution of the Monasteries in the middle of the sixteenth century the balance of supply and demand was affected, especially by the sale of timber from newly acquired monastic land to help pay for the purchase, so that it was necessary for Royal Proclamations and Acts of Parliament to attempt to re-establish the balance later in that century. Although there were many complaints and warnings some sort of balance was indeed restored until the greatly increased demand for charcoal in the late eighteenth century, and for ship's timber in the nineteenth century, meant that the supply of home-produced timber was no longer adequate.

The pre-eminent timber for all purposes was oak. Such trees flourished in most parts of England and Wales: 'to endure all seasons of the weather there is no wood comparable with it', wrote Neve in 1726. Straight oak trunks made posts and beams, thinner branches made rafters and joists, curved boughs made braces, sawn sections made floorboards, inferior stuff when split made laths and wattles; the bark was essential for tanning, the twigs made a merry fire. It is hardly surprising that the sturdy oak remains part of our folk lore. To a very large extent the study of traditional carpentry is the study of the use of oak.

Elm was an alternative for some purposes—for floorboards for instance—but was not commonly used in buildings before the eighteenth century. Elm was reckoned to be most reliable when kept permanently wet—as when used in piling—or permanently dry—as when used in beams kept entirely within the building. Elm was tough and in demand by various wrights for making wheels and machinery: one would expect a mill to incorporate some elm wood.

Ash and chestnut were used in building and to an increasing extent as oak became more scarce and expensive. Chestnut was durable like oak but ash was said to be susceptible to damp. Beech was used in joinery but less often as a building timber. Poplar was used for ancillary purposes as for scaffolding poles, for instance, but this timber has also been recognised in use for cruck blades in buildings in Herefordshire and Worcestershire. White, grey and black poplars are native to this country and as very fast-growing trees could help fill any gap caused through scarcity of oak. (2)

Scarcity or inadequacy of native timber meant ever-increasing use of foreign timber, especially softwood. Native oak produced timbers which were relatively short and rarely completely straight and so the carpenters working on the great buildings of the medieval period had to search long and far to discover suitable timber. Fir, known as deal, was imported on ships and by floating from the Baltic and was denominated 'Riga, Memel, Dantzig etc. fir' according to its port of origin. The use of deal has been recorded from as early as the thirteenth century but its use did not become common until the late seventeenth century nor did deal supplant oak to any great extent until a century later. From the late seventeenth century onwards deal was in demand for floorboards, and building accounts of the eighteenth century often refer to the carriage of seasoned deal from ports all round the coast. The stability, uniformity and ease of working of deal also encouraged its use for staircases, panelling, window frames and other joinery work.

Once the timber trees had been selected it was customary to fell them during the winter so that the green timber was easily worked during the building season which began in spring. However, the bark was most easily peeled if the tree was felled when the sap was rising and so, at least from the early seventeenth century, it became the practice to fell in the spring. Bark was much in demand by tanners and could be worth as much as the growing timber itself. The notebooks of Sir Roger Pratt, prepared during the middle years of the seventeenth century, point to the complicated economy of timber conversion: the trees having been bought as growing timber, the bark would more than pay for the felling, the lop and top i.e. the thinner and otherwise useless branches, when sold, would pay for their own removal and usually for carriage as far as twenty miles, the chips resulting from trimming would pay for the cost of trimming and, if the timber was to be squared, the slabs of rounded section would pay for the sawing. Obviously nothing was wasted and the carpenter was able to work his raw material virtually free from the costs involved in changing its state from a growing tree to an inanimate building material.

At least he was so favoured if the wood or coppice was easily accessible. Transport was the unknown but considerable cost. The further into the forest or woodland the carpenter had to penetrate in order to find suitable material the greater was the cost of dragging the timber to the road, track or waterway. The longer the journey to the building site the greater the cost of hauling logs or timber on pairs of wheels with the use of great teams of oxen. The more preliminary work that could be done at the felling site the less the load to be transported, but the greater the degree of conversion the greater the danger of damage in transit.

Nowadays we are able to use machinery and glues to take natural timber to pieces and put it together again to our own requirements. Before machinery was available and when tools were of indifferent quality and labour slow and expensive, the carpenter had to match parts of trees as well as whole trees to his requirements. The detailed study by Rackham of a sixteenth-century

2. Cholstrey Court Barn, Leominster, Herefordshire
The cruck trusses are made of poplar.

farmhouse in West Suffolk has shown that no fewer than 330 separate trees were felled and used in its construction. Most of these were very small and used for rafters, joists or studs and only a few were large and uniform for use as tie-beams, posts or wall-plates. We assume that four fair faces are needed and that all timbers have to be straight in all directions; actual buildings show that many irregularities can be accommodated if the design is so arranged, for instance, that rafters and floor-joists may be curved on plan provided they are straight on section.

Structural timber was generally used unseasoned. Green timber was easier to work and so many timbers were used in a largely natural state. Until recently most buildings were very well ventilated and so the difference in atmospheric conditions to which the inner and outer were exposed was much less than at present. Only thin pieces such as were needed in wall panelling, doors, window frames and staircases—joiner's work in fact—called for seasoned timber except that within the carpenter's province the floorboards could not be made satisfactorily from green timber. The period for felling to erection was short; much capital might be tied up in growing timber but that was an appreciating asset for the landowner; it was important that little capital was tied up on the slow drying or weathering of felled timber.

There was a limit to the extent that growing timber could match intended use and some conversion was usually necessary. First the pieces had to be squared for many of the intended uses and so the rounded slabs were sawn off leaving as much timber untouched as possible, chamfers or waney edges remaining between the sawcuts of the slabs so as to be moulded later or left in place if the piece was not destined for a prominent position. A piece of square cross-section would suit as a beam or wall-plate or purlin but rectangular sections were needed for many pieces, notwithstanding the waste entailed in changing the shape of a post from rectangular at its head to square in its body. At first, logs were halved by sawing vertically in a sawpit. This might be literally a pit dug into the ground with the log slowly advanced on a cradle as the skilled sawyer on top pulled up and guided the blade of the saw while the pitman down below drew down the saw in its cutting stroke. Alternatively a set of trestles was built whereby the whole operation could be conducted above ground. Mechanical saws powered at first by water and later by steam engines eventually replaced the pitsaw; early models were reciprocating saws in imitation of the action of the pitsaw but from the middle of the nineteenth century the more economical and versatile circular saw came into use, the reciprocating saw remaining popular for sawing joists and planks.

There were different methods of conversion whether hand-powered or machine-powered saws were used: straight up and down sawing produced slabs or planks which were liable to warp; quartering gave timbers of better quality but still liable to distort whereas radially cut members were the most stable of all but involved much waste in the conversion. Where visible the cross-section of a piece of timber will indicate from which part of a log it has been taken while the surface of a timber seen as left from the saw will indicate by its marking what type of saw was used.

The saw having been used to prepare the timber others tools came into action as each piece was shaped for its part in the intended structure. Various lists of carpenters' tools have been published some including items recognisable as tools still in use, some familiar as tools but with names long obsolete, and one or two items oddly named and hard to understand. The tools group themselves into those used for shaping and smoothing the timber, those for setting out the work and placing members in position, those used for making recesses or apertures as needed for jointing, and those used in erection.

For shaping the timber we have already noted the felling axe and the various types of saw: the pit-saw and its mechanical successors and the cross-cut saw for cutting timber to length. Then there are the various axes much used at one time by the skilled workman to make a clean finish to the timber. A special sort of axe was the adze which had a curved blade set at right angles to the handle. The carpenter stood astride the piece of timber chipping away with the adze between his legs. Nowadays adze marks are sometimes equated with crude gouging at the surface of members, it being supposed that they should have the appearance of antiquity. In fact the adze was almost a precision instrument in the right hands, capable of giving a fine finish to timber. As finishing tools the axe and adze were really superseded by the plane. Before machine planing became available, three types of plane were used successively by the carpenter: the jack plane was a long wooden block with a handle and a slot for the blade and the trying plane was similar in arrangement but had a different handle. The smoothing plane was a short rounded or bowed block of wood, again with a slot for the blade but without handles; the smoothing plane was used to remove any slight imperfections left after the work of the other two had been completed.

For setting out the work the carpenter had for centuries to manage without the implements now available. The principle of the dangling weight was used both for the plumb rule (needed to ensure verticality) and the level (needed to allow for horizontal surfaces) until spirit levels came into use. Mortice holes were set out with the aid of a gauge which in its basic form consisted of a block of wood adjustable along a wooden bar carrying a scribing point.

Joints were made with the help of the wimble or auger, the chisel and the tenon saw. A mortice slot was started by boring holes to the required depth either by means of the wimble which was rather like the modern brace and bit or by means of an auger, a cutting head on an iron bar with a handle at the opposite end. The slot was then completed when the remaining wood was cut away by means of a chisel head set in a wooden handle and struck by a wooden mallet. Other chisels lacked the handle and were struck by a hammer. The tenon was formed at the end of a piece of timber with the aid of a tenon saw which had a small serrated, rectangular metal blade attached to a wooden handle. Wimbles and augers were used in boring for other purposes also, such as to make holes to receive staves and for pegs, just as small saws of various shapes were used to make other joints. (d7)

In assembling the various pieces of timber together the carpenter and his helpers made use of wrought-iron crowbars and wooden levers together with levels, a plumb rule and the square so as to ensure the accurate positioning of one member against another. Withdrawable hook pins were used for temporary security at the joints until the final pegs were driven home.

As much assembly as possible was done on the ground. Sometimes sets of timbers such as roof trusses were assembled on a framing ground, adjusted as necessary, taken apart and re-assembled on site ready to be reared or hoisted into position. Otherwise trial and final assembly could take place on site. The outer faces of timbers intended for a wall or those facing into a room as a partition or those forming a roof truss were placed with the best face flat on the ground, any differences between the thickness of members in the assembly would thus face into the building, into a passage or into an inferior room. Prefabricated sections of buildings were put together in sizes large enough to simplify assembly but small enough to be manhandled into position. Apart from the use of terms such as 'rearing' or 'raising' there is little documentary evidence of how timber buildings were actually put together. The illustrations to medieval or later writings or the views in the background of paintings of Jesus as a carpenter are rarely convincing as representations of actual carpentry techniques. Close study of buildings, especially as they are being dis-

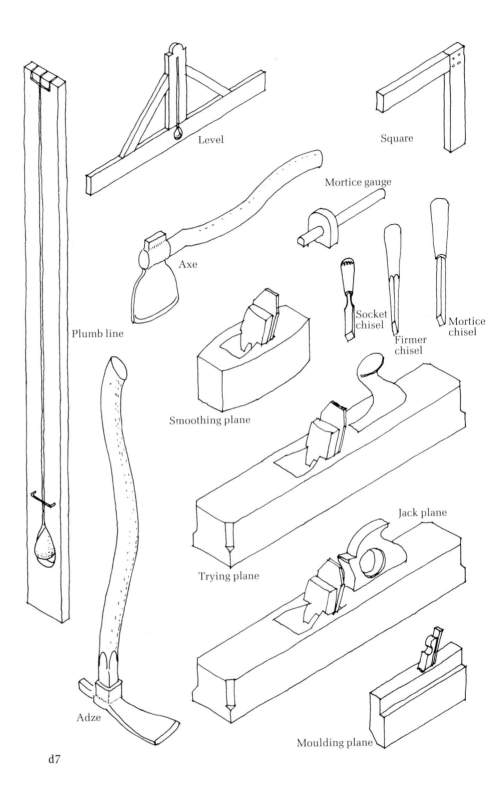

Level

Square

Mortice gauge

Axe

Socket chisel

Firmer chisel

Mortice chisel

Plumb line

Smoothing plane

Jack plane

Trying plane

Adze

Moulding plane

d7

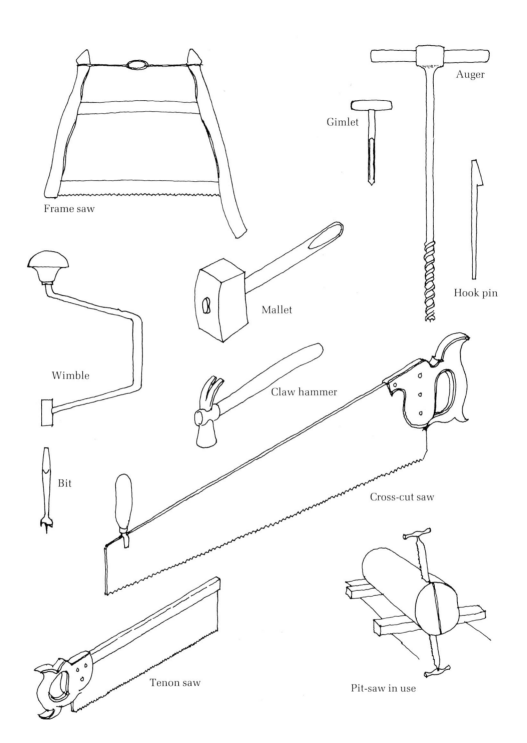

Frame saw

Gimlet

Auger

Hook pin

Wimble

Mallet

Claw hammer

Cross-cut saw

Bit

Tenon saw

Pit-saw in use

mantled has enabled some essays at the rediscovery of construction processes to be made by C. A. Hewett and F. W. B. Charles and others.

These studies have been helped by the practice among carpenters of marking their timbers not (except in rare instances) as a means of identifying the craftsman due for payment as was done by masons, but as a means of identifying members whose joints have been tested, taken apart and subsequently reassembled. Such carpenters' marks are most often found on roof trusses and wall and partition frames, they are sometimes found on floor-joists but rarely on wall-plates. Examples are known from early medieval times and the practice was continued through the nineteenth century as long as timber roof trusses remained in use. Early carpenters' marks were bold, rather sweeping scratches made with a special tool called a race knife, marks of the seventeenth century were more carefully made with a gouge and knife or chisel while those of the eighteenth and nineteenth centuries were small neat and deep marks made with a chisel. (3)

Several systems of marking have been identified. In one the practice was to identify all the members in a truss or cross-frame with a single number, strokes making a Roman numeral on one side of the truss or frame and those same numbers with an extra stroke or tail on the other side. The marks would appear on the upper face as the fair face was laid on the ground for assembly, and in a house the trusses or frames would be numbered in sequence of erection normally starting at the 'upper' or socially superior end. In another system each joint was separately numbered: J. McCann has shown how in a multi-storey building with several tiers of studs the joints were numbered in sequence in each tier but there is an extra symbol to identify the tier.

Generally the marks were based on Roman numerals but with precautions taken so that mistakes were not made when reading the marks upside down. Thus a 4 was always shown IIII while a 6 was the only combination of V and I. Strokes were run together in the larger numbers such as XV or XX. Although the gouge mark C was used it was as a symbol rather than meaning 100, and CCCC used in one part of the building would correspond to IIII in another. Occasionally scratch marks running right across the joint were used in place of numerals. Carpenters' marks make a fascinating but frustrating study: fascinating in the sense of providing clues to be unravelled, frustrating because the marks may have been lost through later decoration of the timbers, because members have been displaced without new marks having been added when timbers were reused, but mostly because the carpenters were not necessarily as thorough or consistent in the marking as we would like.

Carpenters' marks may also be confused with other marks and graffiti. Occasionally marks on timbers make up a genuine date for the building, an unusual example being the reference to the second year of the reign of Edward IV (i.e. 1462) found on a piece of timber built into the gatehouse to Hertford Castle. Sometimes the carpenter signs his work, though rarely as boldly as Richard Dale signed his great bay windows added to Little Moreton Hall, Cheshire, in 1559. Most often there are carefully cut initials and dates representing the idle moments of boys in the nineteenth century.

Much of the timber work of box-frame and post-and-truss construction seems to have been assembled piece by piece, and generally the great frames

3. Churche's Mansion, Nantwich, Cheshire
Incised carpenters' marks are to be seen.

of aisled halls seem to have been put together in a sequence dominated by the need to place arcade plate on arcade post and tie-beam on both. In some forms of construction the finished design seems to have been affected by the wish to assemble the frame on the ground and then raise or rear it into position. Cruck trusses and base cruck assemblies were clearly suitable for rearing and sometimes frames for aisled buildings incorporated a lower tie-beam or straining beam which, being braced back to the posts, allowed such frames to be reared but still completed with wall-plate and tie-beam according to the standard practice of 'normal assembly'. The rearing of cruck-trussed buildings or the raising of frames and separate assemblies in other timber structures required more hands than the carpenter or the building owner could provide. A 'raising bee' was therefore held and was an occasion for the men of a community to get together to help a neighbour, the women to provide a feast and all together to provide a celebration of the creation of the new building. The building owner was expected to provide the drink and references in building accounts to expenditure at the 'raising' show that such hospitality was not cheap. The practice has long been forgotten in this country but remains a part of folk lore in Canada and the USA, was imortalised in the barn raising scene in the musical *Seven Brides for Seven Brothers* and actually remains in use in some communities in Pennsylvania. (d8a, b, c, d)

One further set of operations deserves mention: the alteration, repair and improvement of timber-framed buildings. One alteration frequently made to medieval buildings was the insertion of a floor into an open hall; this required the addition of new members to support the new floor and alteration to existing members. Special joints were required which allowed new items to be added to an already existing frame. Repair has been necessary in the past as well as at present as timbers decayed, not so much from dampness in roof or wall since few roofs had gutters to be blocked, wattle and daub infill did not retain dampness as did brick nogging, and ample ventilation prevented dry rot, but rather from rising damp leading to decay of sills and the feet of posts and studs. Many old timber-framed buildings show the holes through which iron 'needles' were threaded to allow the weight of the building to be supported on props or cradles while the decayed portions were cut out and replaced perhaps at a higher level on a more durable and damp-resistant plinth. Such buildings may also display 'scotch holes' for props used temporarily during construction or repair.

Questions are sometimes asked about foundations for timber-framed buildings. On normal sites foundations in the modern sense were rarely provided: point loads of cruck feet or posts were transferred to padstones or 'stylobats' and thus, with a certain amount of spread, to a bearing layer in the soil. Or the posts and cruck feet and the studs of a timber wall were tenoned into a timber sill which widely spread the loads on to a stone, clay or brick plinth wall. There is some evidence in early prints to suggest that the practice of erecting a frame on a sill carried on low cradles, trueing up the timbers, pegging the joints and then building a plinth up to meet the sill was used in the past as well as being a practice of the present day.

Jointing

Having decided that timber was to be assembled to make a frame for his building, the traditional carpenter was faced with the problem of linking the various pieces of timber one to another by means of joints. Since much structural design seems to have been based on the principle of propping one member by another the problem would not seem to be difficult to solve: one member would simply butt against another. However, rather more than simple propping was involved and over the centuries joints were devised which countered the various ways in which one piece of timber tried to come apart from another.

Among the simplest joints were those concerned with location: one member carrying a load from another had to be located so as best to receive the load. Other simple joints were concerned with overlapping: one member passed over another and it was necessary or desirable that the two should be attached to each other. More complicated were the joints dealing with a tendency to withdrawal: one piece of timber trying to fly away from another. Bearing joints dealt with the circumstances in which one member was carrying another. Lengthening joints were required when one long member was to be made of two shorter members and often it was expected that a uniform cross-section would be maintained. Sometimes joints had to perform several tasks in two or three planes at the same time: these compound joints were correspondingly complicated and the ingenious solutions developed are monuments to the carpenter's craft.

These jointing problems had to be solved with economy of materials and labour. To cut a hole or recess or slot in one member in order to receive another inevitably creates a weakness and carpenters had to design their joints so as to minimise such weakness and, especially, to ensure that timber providing strength in one part of a member was not wasted through weakness of the joint in another. Like whole frames the separate joints were liable to failure through torsion, crushing, bending or shear and had to be designed accordingly.

The joints had to be tight-fitting. A loose joint was liable to come completely apart or, by allowing the jointed members to move, to put unintended loads on some parts of the structure. At the same time, the joint had to be cut, tried, taken apart, adjusted if necessary and then put together again in the finished building and so it could not be too tight. There were various devices to help in tightening the finished joint. One was 'foxtail wedging' whereby thin wedge-shaped pieces of wood were driven into cuts in the end of a tenon so as to tighten a mortice and tenon joint. Another device was 'drawboring' whereby peg-holes in the sides of mortice and tenon were deliberately drilled out of alignment so that the peg, when finally driven home, drew the parts of the joint together. A tight fit was aesthetically pleasing and some medieval buildings retain joints of hairline thickness even after many centuries of building movement. The use of 'secret jointing' whereby one part of the joint was concealed by another was also an aesthetic device much practiced.

Carpentry, like so many other aspects of building, involves a balance between materials and labour, involves a choice which changes from time to time with changes in the relative costs of the two items. When materials are cheap and labour dear it is foolish to spend much labour in order to save on material. Thus cheap and plentiful timber would normally go with simple jointing. When materials are scarce and expensive then labour might be economically used in complicated jointing to ensure the best use of available material. When materials and labour are both expensive then there is an incentive to search for new and cheap materials requiring less skilled and expensive labour. There is little doubt that costs of scarce oak and the expensive hand labour involved in traditional jointing led to the use of cheap softwood timber, cast-iron and wrought-iron members and the elimination of jointing more complicated than what could be done with nails and bolts.

Consideration of locating joints introduces that most versatile of joints, the mortice and tenon. If a stud were to be located in relation to wall-plate and sill, a strut in relation to principal rafter and tie-beam or a brace in relation to post and beam then a mortice and tenon joint was required. In its simplest form the mortice was a slot cut in one member while the tenon was a tongue cut in another member. The tenon was fitted into the mortice, kept tight with wedges and resisted withdrawal through the use of a peg or pin running through the tenon and the sides of the mortice. If one member was narrower than the other then a bare-faced tenon (one with only a single shoulder) was used; if one member was simply to be located on top of another as a crown post on a tie beam, then a short stub-tenon without a peg was used.

An even simpler joint, the lap joint, was used where one member passed over another. This could occur in a single pair of coupled rafters where one rafter was half-lapped over its counterpart at the ridge. It could occur in a scissor rafter roof where the 'scissors' lapped over the common rafters at each end while lapping over each other in the middle. A passing brace lapped over several members in one aisle truss. A concealed down brace might lap over several studs in a jettied wall. In the simplest version of the lap joint neither member was cut and only a peg related one to another; more securely, one member was cut away so as to fit onto another; more secure still was the joint in which one member was half-lapped over another: more secure, but at the cost of weakening both members.

Bearing joints were related to both locating and lapping joints. A floor-joist, for instance, could simply run over the beam which was to carry it but then there would be a tendency for the joist to slide forward, backwards or sideways off the beam. Nowadays we might nail or strap one to another but when nails, straps, bolts and screws were very expensive the solution was to join the joist to the beam by means of a bearing joint. At its simplest the joist was cut away to rest more securely on the beam but it was safer to rest the cutaway end of the joist into a socket cut into the top of the beam. This still relatively crude solution could raise the danger of failure through shear of the joist and through bending of the beam from which a vital part had been cut away. In any case joists resting on a beam made a deeper floor with a nasty gap between each pair of joists passing over the beam. A shallower depth and a less unsightly effect was produced when the joists, cut to require the minimum loss of timber at the end. rested on a ledge cut into the middle of a beam whose valuable top and bottom surfaces were left intact. Where one floor beam met another, as when a binder met a girder in a large floor, and where a uniform top surface was to be maintained, then the ingenious tusk tenon joint was brought into play giving the greatest security with the least loss of essential timber to either one or the other beam.

In spite of every effort the traditional carpenter could not avoid placing some members in tension with the consequent tendency for such members to draw apart from the others to which they were attached. Up to a point a pegged lap joint or a mortice and tenon joint would resist withdrawal but in both cases the resistance depended on the peg which was a thin piece of wood, not necessarily of good quality, distorted by drawboring and not intended to resist withdrawal anyway. At first the notched lap joint was used whereby both the end of one member and the recess cut into another to receive it were shaped to counter withdrawal. This was satisfactory when all was well, but the surface resisting withdrawal was slight and shrinkage or movement of one or both pieces could cause failure. The notched lap joint was succeeded by the dovetail lapped joint, shaped so as to give a longer resistant surface. The single or double-sided dovetail proved a very satisfactory joint to resist withdrawal as long as shrinkage was limited.

Lengthening joints were the most complicated of all. The problem was to make a long piece of timber suitable to serve as a wall-plate, arcade plate or collar purlin out of sections whose length was limited by the timber available. The joint had to maintain a uniform cross-section for the members, to resist bending or sagging as much as possible and to resist sliding—the tendency of one piece to move longitudinally away from another. The scarf joint was devised to deal with this problem. A scarf could not make two pieces of timber as strong as one, but the many variations of shape for each end, together with the use of keys and wedges were testimony to the efforts to approach that ideal. Specially complicated scarf joints were devised in the attempt to produce a tie-beam capable of spanning a wide room usually as part of a queen post roof truss. Long timbers were cut, taken apart and put together again to make the best use of the best sections. Scarce and expensive though the timber might be it was apparently considered worth-while to invest a lot of effort in jointing until eventually wrought iron and then steel became available to do the same job cheaper and better.

The most common type of scarfed joint is the splayed scarf in which one piece of timber is cut to a gentle splay corresponding to another splay on a second piece of timber. To lock the two pieces into position it is usual to have a break or table in the splay and this may be tightened by the insertion of wedges or a key. The edges of the splay are usually splayed or undersquinted at the abutments so that the tightening at the table gives something against which each member can be tightened without sliding away. Alternatively the abutments may be slotted to receive tenons as bridle joints. In another version the abutments are cut away to receive a sally, a pointed shape rather like the prow of a ship. Thus each part in turn of the splayed scarf joint has its task to perform in ensuring a tight joint of maximum strength but maintaining the uniform section of the timbers to be scarfed.

Another fairly common type of scarfed joint is the edge halved joint in which the bottom of one piece of timber is cut away so that it can engage with another member whose top has been cut away. Once again a change in section or tabling helps to provide tightness which is ensured through the use of wedges or keys. Abutments in these joints are usually square but they may be under-squinted to help give an extra tightness and neatness or splayed or over-squinted to give extra ease of assembly. Alternatively the abutments may be bridled or sallied. Rather similar is the face halved scarf joint in which one cutaway member slides vertically over a member correspondingly treated. In a refinement the abutments are cut away in a sort of counter-shape as bladed abutments. Both edge halved and face halved scarf joints depended on pegs or, later, bolts for tightness and completeness.

There are other types of scarf joint, some of great complexity, but the main alternative method of lengthening members was with the aid of a fished joint. In the simple fished joint a free tenon was secured in two open-top mortices and then pegged into position; this could be achieved without altering the cross-section of the elongated timber. Fishplate joints entailed the use of pieces of wood (or later iron or steel) on each side or top and bottom of the members to be joined and a large number of pegs or bolts. Occasionally wooden fish-plates were cut so as to be received by housings in the main timbers or secured by wedges or keys pushing the members and fishplates against the bolts which hold them together.

A further type of lengthening joint was the scissors joint. This could be used as a form of scarfing for horizontal timbers but mostly it was used to lengthen or repair vertical timbers, such as posts. Two opposing splays were cut from one piece of timber and they engaged in a corresponding pair of splays in the other piece of timber while, at least in posts, the load on the posts kept the parts together through compression.

There are few compound joints. By far the most used was that at the junction of post, wall-plate and tie-beam in a box-framed or post-and-trussed building. At the head of the post the wall-plate and tie-beam had to be separately located and the tendency of the wall-plate to slide away from the post or the tie-beam to slide away or topple over both had to be resisted. All this was done by means of a joint at the enlarged head of the post which provided tenons running in two opposite directions and a dovetail lapped joint in addition, all meeting in mortices and halvings to provide a secure joint at this critical position. (d307)

Over the past twenty or thirty years a great deal has been discovered about the development of carpentry joints but there is still not a consensus of opinion on the extent to which jointing techniques can be a reliable aid to the dating of a building, however useful an aid they may be to its understanding. Certainly some joints were introduced, became obsolete and were superseded by others as Hewett has shown, but carpenters from early medieval times at least were very versatile and seemed to have a large repertoire of jointing techniques available to bring out and use as circumstances required. At the same time there was then as now a great difference in resources available for building and a carpenter might use a cheap and simple joint to do adequately on a cottage what he felt required a complicated, expensive, but aesthetically pleasing joint on a church. There was a long overlap between the introduction of one technique and the abandonment of another. It is possible that there were more regional variations than have yet been recognised: research has been patchy. Essex has been very comprehensively studied, other counties rather less thoroughly, some counties have a few buildings which have been intensively studied, other counties have had little systematic study of timber-frame construction whether in jointing or in any other part of the technique.

Cruck construction

Fascination with cruck construction has grown with the study of timber building generally. Perhaps it is the sheer simplicity of the technique, perhaps its improbability, perhaps it is simply that those pioneers of the study of vernacular architecture, Addy and Innocent, were based in West Yorkshire, where examples abound, which has led to this fascination. For whatever the reason the study of cruck construction has commanded an interest out of proportion to the number of examples—though at well over 3000 at the last count they are not few.

Cruck construction makes use of inclined pairs of heavy, usually slightly curved, beams joined together by a collar or tie-beam so as to make an A shape and spaced out in bays so as to transmit roof loads collected from ridge purlin, side purlins and wall-plates direct to the ground. In true cruck construction it seems to have been assumed that the ridge purlin was an important load-bearing member and that its load should be carried down to the ground without placing any reliance on any walling material. (d9a)

The classic cruck truss consists of slightly curved cruck blades carrying a ridge purlin at the apex and the wall-plates on an extended tie-beam. Such a truss was used in a barn or any other building, domestic or otherwise, where a fairly low-set tie-beam would not be inconvenient or aesthetically undesirable. The open truss had a collar in substitute for the tie-beam and the wall-plates were carried by cruck spurs, short cantilevered members representing the projected ends of the missing tie-beam. Open trusses were used in rooms open to the roof where a tie-beam would be awkward or unsightly. A truncated truss had blades terminating at collar level so that there was usually a collar and tie-beam but no means of supporting the ridge. Such trusses were used at the ends of buildings in which a thatched roof covering was to be swept round as a half-hip. (4)

Normally cruck trusses of all three types rose from the ground, ignoring the wall, but in the version known as the raised cruck any type may have risen from within the wall, it being assumed that part of the wall was con-

4. Barn at Corrimony, Inverness-shire, Scotland
The cruck trusses may be seen at bay intervals along the length of the building. Unusual curved yokes join the ends of the cruck blades. Spur ties link the blades to the stone walls. There is a modern roof covering with its own purlins and principals.

5. Tithe barn at Bradford on Avon, Wiltshire
This magnificent roof illustrates various aspects of roof construction. The open trusses are of base cruck form but raised on templates in the stone walls high above floor levels. The base crucks are arch-braced. The various levels of purlins are braced back to the trusses.

sidered strong enough to carry part of the load, rather in the sense of a pedestal stabilised by the wall on each side. A raised cruck altered the proportions of a room, increased the capacity of a barn and might allow the use of the tie-beam in the strong and economical full cruck form.

The base cruck is quite different from the other types and there is some dispute as to whether base crucks have been properly classified and should be called crucks at all. The blades of a base cruck are curved and inclined inwards but end at a collar beam and have some quite different form of construction above. Base crucks usually rise from ground level but may be raised on a wall. They are usually associated with wide spans, were found in buildings of some quality and suggest early rather than late dates. (**5**)

The upper cruck eliminated one of the major features of the full cruck. The cruck blades were curved, inclined inwards and carried a ridge purlin and side purlins, but the feet were tenoned into the tie-beam which was in turn supported by the wall. Thus the feet of the blades did not go down to the ground and the roof loads were imposed on the wall. Upper crucks are associated with the attic floors of houses, lofts in farm buildings and the roofs of mills, warehouses and minor industrial buildings generally: they occupy inferior positions and represent late dates. Often the curve of an upper cruck was not uniform, instead the upper part of the blade was nearly straight and the lower part curved sharply so that a tenon at its end fitted into a mortice in the tie-beam. There is some possible confusion between this form obviously related to the cruck tradition and a similar form, recommended in the late seventeenth- and eighteenth-century pattern books, for the feet of principal rafters derived from another carpentry tradition.

The jointed cruck solved the problem of creating the desired curve by making each blade out of two parts. The part of the blade which served as a principal rafter was slightly curved, it was then joined in some way to another part which served as a post and whose head was curved more steeply to produce a sharp elbow and whose foot continued down to the base of the wall. The joint could be a simple halved joint, the two pieces faced together and pegged; one member could be tenoned into another, again with face pegs; or the upper member could rest on the lower, securely pegged into position. Jointed crucks are usually associated with small buildings such as houses and farm buildings, with roofs of relatively short span; they are widely but thinly scattered in the north and west of Scotland and in Northern Ireland but are also thickly concentrated in parts of Wales and the West Country. Jointed crucks are most often associated with solid walls as if stone or clay was important for their stability.

In all types of cruck construction there was a choice of apex termination. Given that it was considered important to carry a heavy ridge purlin (even though experience must have shown that in all but the smallest roofs which lacked side purlins the ridge purlin was carrying virtually no superimposed load) some care was taken in this choice. One method was to cross the ends of the two cruck blades with a halved joint and so produce a cradle into which the ridge purlin could sit at an angle. Another method was to halve or tenon the ends of the blades into a short horizontal member called a yoke, the square-set ridge purlin resting directly on the yoke or on a short post rising from it. A third method was to form a cradle out of the ends of the blades, linked by a yoke or a high collar and rest the ridge purlin, flat or inclined, in that cradle. Many other variations have been found but most fall into one or other of these categories. Sometimes each truss in a building may make use of a different termination according to the shape of the timber available for the blades of the truss.

Joints in cruck trusses were very simple. Tie-beams were usually halved into the blades with the extra refinement of a dovetail as if it seemed important to prevent the spread of the feet of the cruck blades. Collars were usually halved with or without a dovetail but could be tenoned into the blades. The simplicity, almost crudity of jointing in even the most refined cruck trusses contrasts with the elaboration of jointing in the simplest frames in other types of construction.

Cruck blades may be almost straight in direction though tapered in width, gently curved or with a pronounced elbow, or have an ogee or double curve from a near vertical base to a near vertical apex. They might be taken from separate trees and only roughly match each other or, more commonly, they would consist of matching pairs sawn from the same tree. They were usually thick but plank-like in cross-section though those taken from separate trees were rounded or only roughly squared, and late and crude examples might scarcely have been stripped of bark. Where each cruck blade rose from a separate padstone then there was usually a taper from a thick base to a slender head, but where the blades rose from a wooden sill the feet of the blades were pared down to meet the sill and so the whole blade was given quite a graceful boomerang shape. In any case the foot of the blade would be shaped where necessary to receive the post or stud of a timber wall. At one time it was thought that there was a development in shape from early straight cruck blades through elbowed blades to late doubly curving blades, but later research has found this belief to be not proven.

The cruck trusses which marked the bays of a cruck-framed building were assembled on the ground and reared into position. Experiments in re-erecting cruck trusses as at the Ryedale Folk Museum, Hutton-le-Hole in North Yorkshire have shown that heavy as the trusses might be, only half the weight had to be lifted at one time and comparatively few men were needed. First the truss was assembled with feet as close as possible to their intended final position. The hardest task was then to raise the head of the truss to shoulder height. While part of the rearing team held or propped the truss at this height others set up short poles ready for the second lift and longer pikes for the third lift. A final haul on a rope threaded through a hole in the collar or tied to the yoke brought the truss upright while other ropes acted as guys to prevent the truss from falling forwards. It only remained to 'walk' the truss one foot after another into the final position on the padstones set for the purpose. Some cruck blades have holes or slots about three feet (or nearly 1 m.) above the foot and these slots may have been intended for the insertion of an iron crowbar or wooden lever gripped by several men on each side for this final manoeuvre, but the holes may equally have been intended for helping the initial lift of apex from the ground, or for needles and temporary supports while decayed cruck feet were sawn off and higher padstones inserted.

Once the trusses were in position they were stabilised by the weight of the ridge purlin, side purlins and wall-plates. In addition, in many examples, especially the larger buildings, wind braces were introduced, rising from the backs of the blades to the side purlins, to provide triangulation as well as reducing unsupported length. The braces also prevented longitudinal collapse while the A-shape of the truss prevented sideways collapse.

In an ideal cruck truss the slope of the blades related by way of the depth of the purlins and the position of the wall-plates to the pitch required for the roof covering. This ideal was rarely achieved. If the cruck blade was too gently sloping the purlin had to be deeply trenched into the upper edge of the blade. If, as was more common, the slope of the blade was too steep and the roof

d8 Cruck rearing

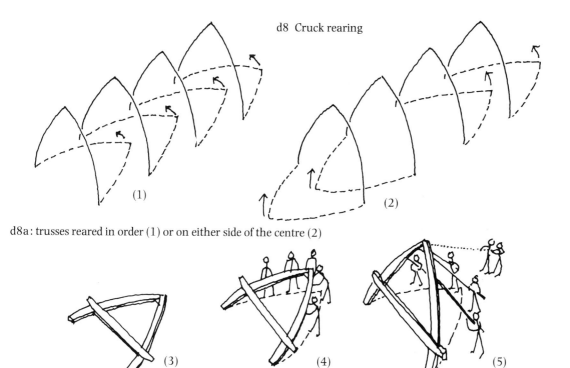

(1)

(2)

d8a: trusses reared in order (1) or on either side of the centre (2)

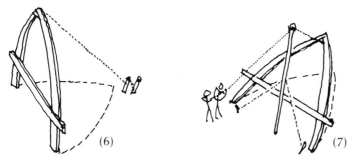

(3)

(4)

(5)

d8b: sequence of rearing from the flat (3) to the first lift (4), the second lift with poles or pikes (5) to the truss tethered in position (6); in large trusses a derrick may help in rearing (7)

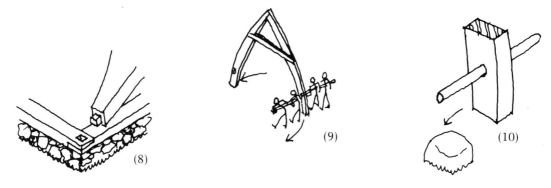

(6)

(7)

d8c: truss feet placed in position: stub tenon dropped into a mortice of a sill (8); feet 'walked' alternately (9) and a foot raised to drop onto a padstone (10)

(8)

(9)

(10)

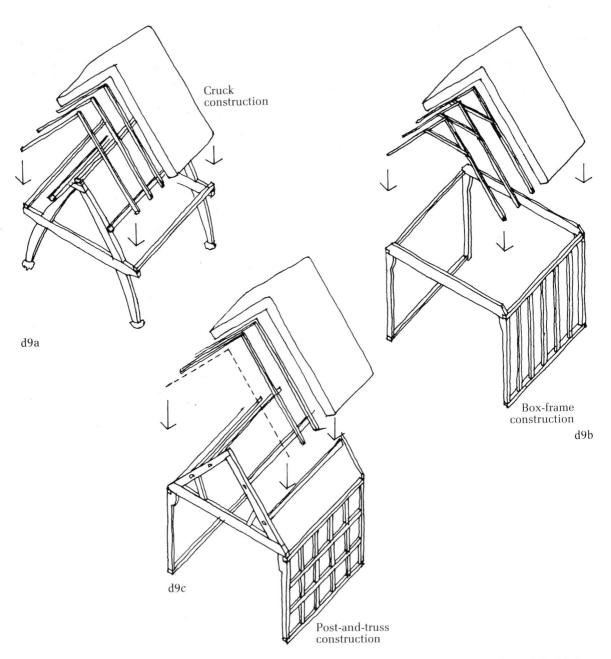

Cruck
construction

d9a

Box-frame
construction

d9b

d9c

Post-and-truss
construction

pitch more shallow the purlins had to be carried on the surface of the blades, set against blocks. Or an extra member as a packing piece on the blade, or running between blade and extended tie-beam had to be introduced.

The concept of the cruck truss required nothing of the wall of a building other than that it support its own weight, and cruck trusses have been found associated with walls of timber-frame, mud and stud, solid clay, stone of poor and good quality and fairly certainly crucks were formerly used with walls of turf. Timber-framed walls were of conventional construction with studs

6. Cruck-framed house at Torthorwald, Dumfriesshire, Scotland
Only poor-quality timber was available for this building and all the members are correspondingly irregular.

and rails forming panels between the structural crucks and between sill and wall-plate; the cruck spurs linked the wall-plate and thus the walls to the cruck truss framework, and other short spurs or cruck ties might provide further linkage between vertical post and inclined blade. Mud and stud walls were like huge panels with staves running from sill or base to wall-plate and thickly daubed with mud on each side. Walls of earth in its various forms such as cob, clay, wichert etc. are found with crucks in many parts of England and Wales. Masonry walls may be contemporary with the crucks but more often they were replacements for earlier and inferior walls. Many a carefully built stone wall in Yorkshire or Derbyshire conceals a set of cruck trusses designed to be clad in some other material. (6)

Timber-frame walling

All walls of timber-framed buildings are there in order to enclose space, to keep out wind and rain and to keep in heat, some walls also exist to keep up the roof and many walls are expected to present a fashionable and decora-

7. House at Pembridge, Herefordshire
This timber-framed house has square panels in the walls and diamond panels in the gables. One end of the building is jettied.

tive appearance to the observer. Timber-framed walls include those which rise in a uniform manner through one or several storeys to eaves level and also those which have projecting or jettied stages rising level by level to eaves or gable verge. In constructional terms, timber-framed walls include those which are divided into structural bays, studs and rails forming panels between the structural members and also those less frequently found in which studs and rails alone provide the supporting structure.

In the simplest box-framed structures a series of corner and intermediate posts rise from sill to wall-plate to indicate a series of bays which are given further definition by tie-beams running at each end and at the intervals set by the intermediate posts. Such a structure would be braced in one of several ways and then the bays would be filled with panels or clad with some overall weather-protective covering. (d9b,c)

Usually the corner posts rise from a plate or sill formed of wooden members running horizontally round the building, halved at the corners and with the corner posts set in stub-tenons on the halvings. The intermediate posts also are stub-tenoned into the sill which may itself be scarfed or made up of several pieces in order to provide sufficient length. The sill is set on a plinth of stone, flint or brick, rising a foot or two from the ground except in late buildings when the plinth might be several feet high. A small number of examples have been discovered, mainly in the North of England of the 'interrupted sill' whereby posts rose from padstones and the sill was more of a long rail running along quite a high plinth and tenoned into the sides of the posts. The sill was a vulnerable member, susceptible to damp rising through the plinth or splashing from the rainwater which fell from gutterless roofs to the foot of the wall, and in many cases the sill has been renewed, or removed to be replaced by a new plinth. In so-called half-timber construction the walls between the structural members were divided into panels and these fall into two main groups: tall, narrow, storey height panels and square, or nearly square panels of which there were two, but sometimes three to each storey height. The tall panels were usually defined by studs only, though sometimes there were light rails running at about half-storey height. The square panels of studs and rails might be divided by subsidiary pieces of timber to form some decorative pattern. The panels between studs and rails were infilled in one of several ways: probably the most common and best-known infill was wattle and daub but this was mostly used with square panels. Tall panels were filled by other materials such as pieces of stone or tile and then daubed or were filled by daub on staves; there are occasional survivors of the butt-edged wooden board fitted between slots to make a panel infill. (7)

Tall panels were often very narrow. In many examples, and especially in early buildings, the panels were barely wider than the studs themselves and in such cases literally half the wall was of timber. With such prodigal use of a valuable material it is hardly surprising that narrow panels are associated with high quality building for clients of high status, and early examples survive because they are part of expensive, durably constructed buildings. The narrow panels between the closely-spaced studs were often filled with small slabs of stone or tile cut to size or with thin bricks made for the purpose and called by contemporaries 'wall-tiles', the various materials being covered with a thin daub of clay, dung, straw and hair mixture inside and out and then finished with a limewash coating. Alternatively, the daub was applied to a reinforcement of thin pieces of riven oak sprung between grooves let into the sides of the studs or of a zig-zag pattern of short staves forced between auger holes on one side and a groove on the other. The alternation of stud and panel along the face of the building gave a rich and decorative effect. (8) (9)

Square panels (rarely exactly square) were almost invariably filled with daub on wattle and staves. The staves, looking rather like the palings of a chestnut fence, were sharpened to a point at one end and a blunt chisel shape at the other. The pointed ends were placed in auger holes bored on the underside of rail or wall-plate, the chisel ends were placed in a groove cut into the top of rail or sill and then sprung roughly into position. Staves were placed some five or six inches apart (about 125 to 150 mm) and a typical panel would have six or seven staves. If the panel was rather taller than it was wide the staves would be placed horizontally but sprung into position in the same way. In the usual arrangement the staves then provided the warp for a weft of wattlework. Split oak rivings or hazel withies were chiefly used for wattling, the one being more expensive but longer lasting than the other. In solid panels the basketwork of staves and wattles was daubed on both sides and then finished with a limewash coating so as to leave the timberwork standing a quarter of an inch or so proud of the panels. Where ventilation was required, as in a barn, the staves were grouped in pairs for strength and a tight weave of riven oak was worked around them but there was no daubing, the crops thus received protection and ventilation at the same time. Nowadays many panels are filled with bricks and a few are filled with stones or flints; nearly always these are replacements. Auger holes and grooves prove that wattle and daub was intended; only smooth-sided uncut timber members indicate original solid filling. (10)

Although the many joints of a timber-walled structure went a long way to provide lateral and longitudinal stability, in practice braces were required to ensure stability through triangulation and in addition some braces gave temporary support during assembly of the frame. Braces have been classified as upward, downward, tension and concealed. (11)

Upward braces rose from a vertical member to a horizontal member, thus a knee brace joining a post to a wall-plate is an example of upward bracing. Early braces of this sort tended to be long and curved, later examples were short and straight. There was little need for such braces to help reduce unsupported length since some support to the wall-plates or the tie-beam in the gable wall of a building was given by the studs. A short straight knee brace gave excellent triangulation.

Down braces fell from a vertical member such as a post to a horizontal member such as a sill, thus a brace running between a sill and a corner post is an example of a down brace. Such a member would help to provide triangulation though not where it was most effective, and whether straight or curved it is easier to see a down brace as locating or securing a post before other parts of the frame were put into position.

Tension, or so-called Kentish bracing, ran between a post and a sill and was virtually always curved. In its standard form pairs of braces linked post

8. Barn near Dilwyn, Herefordshire
Rather heavy members divide the wall into square panels. The upper range of panels has an infilling of wattles woven around staves. The infill of the lower ranges has been removed and replaced by weather-boarding.

9. Cottage at Collier's End, Hertfordshire
The lath and plaster and weather-boarding have been stripped from this building to reveal the main horizontal and vertical members and the lighter studs dividing the walls into tall, narrow, panels. The roof consists of common rafters linked by collars but the lack of longitudinal restraint has resulted in some racking of the roof timbers. The distinction between the box-frame of the walls and the rafter 'lid' of the roof may clearly be seen.

10. House at Benington, Hertfordshire
Close-studding makes for tall and narrow panels in this timber-framed wall.

11. Weaver's House, Stratford St. Mary, Suffolk
The tall narrow panels have infilling of wattle and daub. Bracing is concealed except where the long straight braces link posts and studs.

to sill in each storey of a jettied wall. A somewhat similar pattern may be seen in Suffolk and Essex, for instance, except that the curved brace runs between post and stud and may be related to the idea of the passing brace. It is hard to understand why a bracing member should run between a structural and a non-structural timber, and it may be that such bracing was intended more to help with prefabrication, keeping a set of studs in position until they were placed in the building.

Concealed bracing was used especially in Kent, Sussex, Essex and East Anglia to avoid altering the visible pattern of close-studded walls while still providing the necessary bracing. Concealed bracing as down bracing between corner post and sill or girding beam may be found inside some houses in these counties.

Window and door openings were arranged within the walls as required by the plan and the desired elevational appearance and studs and rails were

12. House at Walkern, Hertfordshire
The timber-framed walls of this house are concealed beneath a plaster cladding.

adjusted in position to provide jambs, heads and sills for such openings. Often the edges were chamfered and moulded out of the solid timber and such clues may indicate the former position of doors and windows lost in subsequent alterations.

As an alternative and eventually as a successor to half-timbering, buildings were clad overall in a variety of materials. Plaster, tiles, slate and weatherboarding were used, separately as a rule but sometimes combined in the same building. Existing buildings were clad either to improve weather-resistance or to provide a more fashionable, more decorative appearance, concealing parts which had decayed or appeared old-fashioned. New buildings were clad rather than left half-timbered for much the same reasons: the clad buildings were warmer, drier, required less maintenance or gave no obvious sign that the structure was old-fashioned; but in addition, new buildings could make use of timber of less than the best quality if it were hidden behind a cladding.

13a. Market Street, Lewes, Sussex
Tile-hanging between buildings clad in mathematical tiles.

13b. The Old Sun Inn, Saffron Walden, Essex
The plaster cladding of the timber-framed walls is moulded and decorated in the technique known as pargetting.

In towns, cladding in tile or slate or even plaster helped increase fire resistance at a time when overcrowding and the warning of the Great Fires of London and other cities and towns made fire resistance an important factor in design.

Plaster cladding consists of a layer of two or three coats of lime plaster applied to riven laths or withies nailed or otherwise secured to the studs of the building. The plaster could be left plain or colour-washed or lined in imitation of painted stone walling, or pargetted i.e. scraped or combed into patterns or moulded in fanciful designs. (12, 13) Plain tile hanging consists of roofing tiles, squared or shaped, hung and pegged to laths nailed to the timber-framed wall structure. Shapes such as fish-scale or scallop, and colour in diaper patterns relieved the plain tile cladding. Mathematical tile hanging consists of tiles rather like plain tiles in size but shaped in section so as to expose a surface rather like that of a brick. When pointed a wall of mathematical tiles gave a very convincing impression of a properly bonded brick wall. (49) Colours for mathematical tiles ranged from a buff, almost like so-called 'white' brick, to a glazed black. The tiles were nailed to laths or against boarding but set in a bedding of mortar. Special corner tiles were produced for use at window reveals as well as at the corners of buildings; special tiles were also scored in imitation of voussoirs to set over windows and door openings. Slate hanging was similar in principle to plain tile hanging. In utilitarian circumstances simple roofing slates were hung on laths to protect the timber-framed walls from driving rain, but otherwise small slates were cut in patterns to give a neat and decorative appearance to a timber-framed wall or a timber-fronted building. Weather-boarding was used in both decorative and utilitarian circumstances. Weather-boarding, either stained or tarred or left untreated was a relatively cheap way of improving the finish and reducing the maintenance on farm buildings or minor industrial buildings. Weather-boarding, either shaped or painted in imitation of stone or simply applied and painted, made for a neat finish and up-to-date appearance to a timber-framed building. Weather-boarding became a reasonable choice of cladding material once softwood timber was widely available and mechanised sawmills could cheaply convert it into feather-edged weather-board sections. (14)

Cladding is mostly a fashion of the eighteenth and nineteenth centuries. Admittedly plaster cladding and pargetting is a long-established technique and some of the famous pargetted buildings, such as the Ancient House at Clare in Suffolk, display patterns which seem to belong to the sixteenth century but general use of pargetting and other plaster claddings came later. The earliest known examples of mathematical tile cladding is on a house at Westcott in Surrey where a tile marked 1724 has been discovered, but most examples are from the late eighteenth century or the first half of the nineteenth century.

Plaster cladding and pargetting are most commonly to be found in Norfolk, Suffolk, Essex, Kent and parts of adjacent counties; tile hanging is found in Sussex, Essex, Kent and Surrey mainly; mathematical tile hanging may be found in Sussex, Kent and Surrey but urban Hampshire and Wiltshire have provided many examples. Weather-boarding is mainly a Kentish technique but many examples can be seen in Essex, Sussex and other counties of South-Eastern England and weather-boarded farm buildings can be seen wherever

14. House at Collier's End, Hertfordshire
The front of this house is covered in a cladding of horizontal weatherboarding.

timber-framed construction remained long in use—in Shropshire and Herefordshire, for example. Slate hanging was popular in Devonshire but examples in the Lake Counties are few probably because even in the towns the tradition of timber-framed building did not linger.

Jettying provides one of the most familiar yet still most mysterious varieties of timber wall construction. (15) The external jetty—whereby the wall of one storey projects over the wall of another—is the most common, but jettying may occur within a building. The external jetty may project to the front of a building, to the sides, to the back or to some combination of all three. The jetty may be of part of a building as in the wings of a Wealden or H-shaped house, or it may be along the whole of the building as in 'continuous jetty' buildings, or in the projecting storeys along the alleyway of a burgage plot in a town. Rather less common, though still often seen, are the jettied gables whereby the triangular gable section of a building—or of a dormer window— projects. Less common still and seldom seen are the internal jetties whereby the wall of a first floor chamber projects into the open hall of a medieval or sixteenth-century house.

The process of jettying is quite simple though the jointing involved may

15. The Chantry, Sudbury, Suffolk
Two adjacent sides of this building are jettied. The arrangement of projecting joists and the sill, posts and studs of the jettied upper floor may be seen.

be complicated. Usually the posts and studs of a lower storey are finished in a beam or plate, the floor-joists of the next storey are projected a foot or two forwards (300 or 500 mm) bearing on this plate, and a further beam or plate is carried on the ends of the projecting joists acting as a sill for the posts and studs of the next storey. This arrangement was often continued storey by storey to the full height of the building. A less common form of construction involved the posts and main floor beams only; pairs of beams project over the posts and the upper part is jointed to the projected beam ends; studs rise from a sill spanning between the posts. Less common still is the hewn jetty whereby posts running through two floors were cut back at the intermediate floor level to take two plates, one projecting slightly in front of the other. Where jetties were provided for two adjacent sides the projecting floor-joists run in one of two directions, those near the corner being received by a diagonal beam which in itself was given some extra support by a shaped corner post. (**16**)

The ends of the projecting joists were at least rounded and might be moulded. Often a plank with vine trail decoration and a battlemented cresting was fixed to conceal the ends of the joists. Sometimes, and especially in the West Midlands and the North West of England, the projecting joists were con-

16. The Chantry, Sudbury, Suffolk
This detail of the corner of the building shows the angled floor-joists and the support given by a curved and boldly projecting angle post.

cealed behind a cove of plaster on battens or of shaped timber with panels sunken to receive plaster. Little Moreton Hall in Cheshire is a well-known example of both jettied construction and the various sorts of coves.

The device of jettying was widely used, especially in the sixteenth century, in buildings ranging from the palace to the stable. Jettying is sometimes believed to have been devised for town building and certainly jettying is often found on the fronts of urban buildings as they face into street or market place. It is also found on the ranges which extend down urban plots as lodging rooms in coaching inns or as subsidiary domestic accommodation behind merchants' houses. Free-standing buildings such as market halls may display jetties on all four sides. However, some of the earliest jettying known is found in rural buildings, especially farmhouses of the 'Wealden' type. There are also a few instances of particularly elaborate farm buildings with jettied upper flooring— at Ulnes Walton near Leyland in Lancashire for instance. The great, though long-demolished, palace of Nonsuch in Surrey was a spectacular country house with jettying. Although the main flourish was in the sixteenth century the practice did not die out until the late seventeenth century and there may be later examples.

Many suggestions have been made to explain jettying. For instance jettying has been suggested as a means of increasing the usable floor space in a building, and this it does but at some expense and, in rural buildings at least, it seems an elaborate way of making use of an ample site. Jettying has been suggested as a means of gaining a legal encroachment on the street or market place of a town; this it does but once again the floor space gained seems entirely out of proportion to the cost of gaining it. Constructional arguments have been advanced. Jettying has been suggested as a means of counterbalancing the weight of a wall with the weight of furniture and people in a room but the projection was usually far too small to provide much benefit; furniture is not particularly heavy, even of the sixteenth-century variety, and the floor load of people in a room is so variable and unpredictable that floors have to be designed accordingly and gain little benefit from counterbalancing the walls. Again, most buildings in town and country have far more unjettied walling than jettied: what works for one part of the building could work for another. Jettying has been suggested as a means of reducing 'springiness' in a floor, but again the benefit could apply to very few rooms in a building. Even defensive precedent has been suggested for the practice of jettying and admittedly the 'brattishing' hoardings (built out from battlements to protect defenders leaning out to fire on attackers) were jetties of a sort, and the medieval timber framing on one of the towers at Stokesay Castle in Shropshire (98) may have had some defensive origin, but again the connection is weak. Rather more convincingly the jetty has been seen as a means of using short storey-height posts. Again, this is true, but posts running through two or more storeys were apparently found to be used in the unjettied back walls of buildings whose fronts were jettied.

The most convincing explanation for jettying is that it was primarily decorative, was meant to impress and could provide a field for display. One rarely finds a jetty which was not highly visible from street, alley or market place in the town or to the visitor to a house in the countryside. The idea of the cornice was familiar in classical architecture and the attention paid to this feature especially in the Early Renaissance in Italy may have had some influence in a much changed form in the timber-framed walling of remote parts of the Western World. Whatever the origin and purpose of jettying, whatever the influence of medieval fortification or renaissance or classical architecture, jettying remains one of the most fascinating aspects of timber-framed construction in Britain, in Western Europe and in many places colonized by Europeans.

Roofing

The roof shares with the hearth symbolic significance in the design of a house, or indeed any other building: the hearth provides warmth, the roof provides shelter; one could no more conceive a house without a heat source than a building without a roof covering. Yet just as the hearth or fireplace or inglenook is so much more than a real or symbolic source of heat so a roof is so much more than simply a device for keeping out the rain. Externally it provides the lid or hat which traditionally completes the architectural composition of a building. Internally it provides the culmination of a tall elevating space as in a hall or church. Overall it provides a test of the carpenter's skill and an opportunity to display, openly or in secret, the technique of his craft. (17)

Yet ultimately the form of the roof is largely determined by the nature of the roof covering and for all but the most expensive buildings this has meant the material cheaply and readily available in the locality. The roofing material largely determined the slope or pitch of the roof. Thus the smaller the unit of roofing material the greater the number of gaps into which rain could penetrate and so the steeper the pitch at which the material must be laid. Thatch required a pitch of 50° or so; plain tiles of clay and thin stone tiles required at least 45°, thick slates and stone flags were traditionally laid to a minimum of about 35°; thin slates of large size and uniform dimensions as well as pantiles were laid often to rather less than 30° while only sheets of lead or zinc could be laid to a pitch which was nearly flat. The roofs used in Mediterranean lands are obviously unsuitable for the British climate and lead roofs were usually given several degrees of pitch.

The roofing material also has had some effect on the shape of the roof: hipped roofs are generally unsuitable for low pitch and large or boldly shaped roofing materials (though low-pitched roofs of large slates were used for special aesthetic reasons) and so one rarely sees hipped roofs of stone flags or pantiles; yet reed thatch suggests gabled roofs while the rounded forms of straw thatch suggest hipped or half-hipped roofs. At all times the pitch and shape of the roof were also affected by fashion: at one time the steeply pitched many-gabled roof was fashionable, at another the low-pitched roof hidden behind a parapet was the shape sought after.

Given the pitch and the shape of the roof have been largely determined, the carpenter's main concern was with the span, the distance between supports to the roof construction. Usually these supports were the front and back

17. Mill at Rossett, Denbighshire, Wales
Here the timber framing has been encased in brickwork but the gabled roofs of one wing and a dormer expose some timber work.

walls of the building—and detailed design was affected by the walling material, a thin timber frame or a thick wall of stone or clay. The supports might be the intermediate row of aisle posts or the partition walls making up the rooms of the building might be designed so as to reduce his roofing problems. In any event a series of constructional problems had to be solved in sequence: roof covering was to be carried on laths, they in turn had to be carried on rafters spaced apart according to the load-bearing capacity of the laths; rafters on all but the narrowest roofs had to be linked by braces or carried on purlins; the convenient span of purlins determined the bay spacing or the distance apart of crucks or roof trusses, and pitch, span, use of roof space and available timber largely (though not entirely) determined the form of the roof trusses.

Roofs consisting essentially of rafters are called single or rafter roofs; roofs requiring crucks or trusses are called double roofs.

The simplest type of roof, sometimes called the couple-close roof, consisted of pairs of rafters rising from the wall-plates and meeting at a halved or tenoned joint at the apex of the roof. Such a roof was only suitable for a narrow span and a fairly steep pitch, having a rafter short enough not to sag in the middle under load nor to spread and so force apart the wall-plates. For longer spans and longer rafters some intermediate bracing was needed: both a collar and a pair of intersecting braces or scissor rafters could simultaneously counter the tendencies to sag and to spread. These with collar, scissor braces and the vertical ashlaring used with thick walls could be lined internally with boarding to produce the tunnel or barrel roof.

Such roofs of rafters or trussed rafters had one serious defect: the absence of any effective means of longitudinal restraint other than what might be given by hipped or half-hipped ends. The most common way of dealing with this problem was through the use of the collar purlin (sometimes more correctly called the collar plate). This longitudinal member joined together the collars, kept them at their proper positions and so maintained stability. The collar purlin was supported by short posts called crown posts carried in turn by tie-beams running between timber-frame or masonry walls. The crown post was braced to collar purlin and sometimes to a collar as well; it was propped in position by braces between tie-beam and crown post. Early crown posts or those hidden or in utilitarian buildings were plain and undecorated. Later crown posts where visible were decorated, the upper braces rising like vaulting ribs from carved capitals on shaped shafts.

Another solution to the problem of longitudinal restraint was provided by the clasped purlin. In this technique a pair of rather slender purlins was threaded between rafters and collars; in this sense they were clasped by the rafters and collar. Where organisation of the roof in bays had been achieved the purlins were clasped between a heavier collar and a pair of stout rafters which below collar level were as deep as principal rafters but diminished to the dimensions of common rafters above.

Most roofs of more than very small span were double roofs having trusses carrying purlins which in turn carried the rafters. Cordingley has divided double roofs according to the way in which the purlins were carried and although this distinction is not perfect and hybrid roofs may be seen, it is nevertheless important. The distinction is between butt-purlins (sometimes called tenoned purlins) and through purlins (sometimes called trenched purlins). In both cases the purlins were carried by heavy inclined timbers— either crucks or by roof truss members called variously principal rafters or blades.

Butt-purlins run *between* the principal rafters and were dropped below the upper surface of the principal so that the tops of common rafters carried by the butt-purlins ran uniformly with the tops of the principal rafters. The butt-purlins did not, of course, simply butt against the principals, they had to be supported and this was usually done with the aid of tenons at the ends of the purlins engaging in mortices cut in the principals, runs of purlins being staggered to allow this to be done. Such purlins provided some longitudinal

18. Former house at Tarvin, Cheshire
This roof truss consists of a pair of principal rafters or blades rising from a heavy tie-beam and carrying a ridge purlin and two side purlins with the aid of inclined struts. The common rafters are irregularly shaped on plan but are flat in the plane of the roof slope.

restraint, preventing trusses from collapsing sideways, as well as providing support for common rafters which could be made out of short lengths of timber. The number of butt or through purlins on each side of the roof was determined by the support needed for the rafters: simple roofs and light loads might entail the use of only one purlin on each side, bigger roofs with heavier covering might demand two, three or four purlins on each side. (18)

Through purlins run *across* the principals. They were laid so as to rest on the upper surface of the principal rafter or blade and either laid against a block or laid in a trench on the upper surface. Common rafters were laid on the purlins but made a continuous run quite independent of the principal rafters. Once again the heavy purlins gave longitudinal restraint to prevent sideways collapse of the trusses as well as giving support to the rafters. Purlins could run over several bays or be made of short lengths lapping at the principal, or be scarfed into long lengths.

Although it was possible in theory to have a pair of inclined principals relying on stout walls to prevent spread, in practice this was rarely done and normally there was a tie-beam which gave security against collapse of the wall through buckling as well as combining with principals to form a truss. The simplest truss consisted only of a triangle composed of two principals and a tie-beam, but this was only suitable for short spans. There was a tendency for the principals to sag under the weight of the roof and the tie-beam to sag under its own weight. Angle struts helped support the principals, a collar helped reduce sag and was also useful in assembling the truss. Struts carried on the tie-beam helped increase the tendency to bend (unless there were corresponding struts beneath) but where this was a serious problem the king post or queen post roof trusses were used instead.

The king post roof truss is one of the most fascinating pieces of timber construction since it illustrates a logical, though incorrect, assessment of the forces acting on a truss existing side by side with an apparently unlikely, but in fact quite correct, understanding of what actually happened. Traditionally, carpenters in this country regarded the king post as a prop rising from a tie-beam to carry a ridge purlin and its load. We have seen that the designers of cruck trusses regarded the purlin as an important load-bearing member though in practice on all but the smallest roofs, which lacked side purlins, little load was carried. The ridge purlin was mostly carrying its own weight and of course there were many roofs which had no ridge purlin at all. Nevertheless some carpenters regarded the ridge purlin as important and propped it up by means of a long king post rising from a stub-tenon on top of the tie-beam; sometimes they braced back the ridge purlin to the king post so as to reduce its span and to give some longitudinal stability to the roof. At the same time principal rafters carrying purlins rose from the ends of the tie-beam to meet at the head of the king post which was often shaped to receive them. Because the principal rafters tended to sag they were braced back to the foot of the king post which was also shaped to receive them. In fact the principal rafters were trying to push up the king post while the tie-beam was trying to fall away. The carpenters must have noticed a tendency for this to happen—the joint at the foot of the king post tending to open—because occasionally one sees in traditional work of the late seventeenth and eighteenth centuries a dovetail joint instead of a stub-tenon between the king post and the face of the tie-beam. But eventually the true situation was perceived, and the king post (now described as a suspending member) and the tie-beam were joined together with an iron bolt or strap.

On the continent of Europe the true situation had long been appreciated and seventeenth-century English architects who had learned their architecture abroad, such as Inigo Jones and Sir Christopher Wren, provided for straps to join king post and tie beam in all their trusses.

All the roof trusses so far described depended on the use of the tie-beam, a member which was expensive and hard to obtain in sufficient length for the required span and in sufficient depth to reduce sagging. It was also unsightly in rooms exposed to the roof and in stone-walled buildings, at least, might not be needed for the stability of the wall. One popular method of eliminating the tie-beam and producing an attractive roof structure was through the use of the arch-braced collar beam truss. In this truss, principal rafters rose from the wall heads to the apex of the roof and were connected by a fairly heavy collar, the angles between principals and collar were filled with an arch-shaped brace and the truss was strengthened as well as made more decorative. As a further development long arch-shaped members were introduced to rise from part way down the wall to meet at a point at the bottom of the collar. Sometimes the collar was itself arched or cranked to increase the effect.

As in cruck construction, wind braces (sometimes called sway braces) were used to resist any tendency for the trusses in double roofs to collapse sideways. Where the braces rose from principal to purlin they could help support the load on the purlin; downward sloping wind braces were more decorative and the effect of a combination of both sorts dividing the roof into large quatrefoils could be quite dramatic. In late examples a long narrow brace stretching from near the ridge to near the wall-plate was used, rather as trussed rafters are braced nowadays.

The detail at the junction of roof and wall caused problems which in some cases led to ingenious solutions. Where a timber roof met a timber-framed wall there was no problem, constructional or aesthetic. Where a timber roof met a stone wall, the wall-plate was best positioned near the outer face of the wall leaving an unsightly gap or awkward beam-filling on the inner face of the wall. The most usual solution to this problem was the use of 'ashlaring'. Joists ran between the outer wall-plate and an inner wall-plate to receive studs which dropped from the rafters; visibly the rhythm of the rafters was continued down to the inner head of the wall.

Where the roof space in a multi-storey building was to be used as attic or garret accommodation, the sharp angle between principal rafters and tie-beam reduced effective headroom to a zone in the middle of the roof. Upper cruck construction neatly solved this problem, the upper crucks gracefully bending at the elbow and their feet meeting the ends of the tie-beam at or near right angles. A somewhat similar solution was used in the late seventeenth and eighteenth centuries with the sharply cranked ends to otherwise conventional principal rafters. A more radical solution involved the interrupted tie-beam where a section of tie-beam was tenoned into a vertical post running between floor beam and principal rafter. Another version of the interrupted tie-beam was the 'post and pad' detail where the ends of the principals were let into short timber pads; short timber posts dropped from these pads to a floor beam and were braced back to the end of this beam. Thus were produced neat and quite strong details which may be seen in granary, warehouse or workshop floors in buildings of the late eighteenth and nineteenth centuries.

Wide span roofing

Until the introduction of wrought iron and later steel to their repertoire carpenters struggled with the difficult problem of providing a wide span roof. A long roof presented no real problem, bays were simply added to bays to produce the required length, but many building types required a wide central area: churches, assembly halls, market halls, large barns for instance. In some cases the covered floor area could be interrupted by posts or piers without causing much difficulty; in other cases an unobstructed floor area was required for reasons of use or reasons of appearance. Where intermediate support was acceptable the problem could be solved with the timber available but where an unobstructed span was required then the length of timber available for a tie-beam limited, even dictated, the solution. Aisled buildings gave a wide span but with some interruption to the floor space; base cruck and

19. Barley Barn at Cressing Temple, Essex
The interior of this aisled structure shows the range of arcade posts between nave and aisles carrying the square-set arcade plate with the aid of braces. Tie-beams link the arcade plates above each arcade post though here there is an additional beam below the tie-beam. Long braces run across this beam between arcade post and tie-beam.

hammer beam buildings had no such interruption. Eventually fir timber in long lengths became available and queen post roof trusses allowed great spans as well as uninterrupted floors.

An aisled hall consisted of a conventional timber frame forming the nave with rafters sweeping down one or both sides to cover the aisles and, in the widest aisled halls, to cover double aisles on each side. The essential structure of the nave consisted of posts (known in this case as arcade posts), plates (here known as arcade plates) and tie-beams which were usually braced back to the posts. The arcade plates were set square on top of the posts rather than tilted like purlins and so the joint between plate and tie-beam was just like that at the head of the posts in an unaisled hall. Roof construction above the tie-beam varied but a crown post and collar purlin roof was often used. The line of rafters was continued down to wall-plates carried on a series of lower outer posts rising from sills and which helped to form the wall of the

20. Tithe barn at Great Coxwell, Berkshire
The magnificent stone walls conceal an aisled timber construction within.

building. Horizontal ties ran from, or just below, the heads of these posts to meet the arcade plates and often these were in the reversed assembly form of wall-plate over tie-beam on posts, the positions being changed in normal assembly at the head of the arcade posts. Sometimes there were diagonal braces running between wall posts and arcade post and usually braces ran between arcade post and arcade plate, giving a sort of arched effect and so the name 'arcade' to an assembly which did not in fact include a true arch. In early examples, passing braces were used to tighten the many-jointed frame and help to prevent any danger of sideways collapse; passing braces were long thin members running from wall post to rafter and halved into all the members passed on the way. (**19**) (**20**) (**21**) (**22**)

The aisled hall was an elegant form of timber-frame construction. Roofs were usually steep (though in Yorkshire shallow pitched, stone-flagged aisled halls were built) and an observer at ground level may stand in awe at the

on previous pages
21, 22. Tithe barn at Great Coxwell, Berkshire
The great aisle posts are raised on stone pedestals and carry first the tie-beams and then the arcade plates in reversed assembly. Braces are doubled in both directions. A separate structure gives intermediate support to the purlins in the middle of each bay.

23. Church of St. James and St. Paul, Marton, Cheshire
The interior of this aisled timber-framed church shows very heavy constructional members with deep and curved braces between posts and arcade plate to give a truly arcaded effect.

forest of timber members, heavy and light, rising in the gloom above him. Examples are concentrated in the eastern and south-eastern counties of England, though there is a lesser concentration in West Yorkshire and there are outliers over the Pennines in East Lancashire. From early medieval beginnings the aisled hall type continued to be used in barns certainly until the late seventeenth century and probably into the eighteenth. (23) (24)

Base cruck halls are far less numerous, less widespread in their distribution and were used over a shorter period and for a more restricted range of building types. A base cruck consisted of a pair of inclined curved members rising from well below the eaves of a building and meeting a heavy horizontal member at collar level. Various sorts of roof construction might rise above this collar beam but they usually incorporated horizontal plates running across the collars from truss to truss. In that they were inclined and curved, or elbowed, these have come to be called base crucks but in that they do not attempt to carry a ridge they might equally have been considered as posts, inclined to meet a shortened tie-beam and with reversed assembly. The construction above the collar level tends to suggest an association with crucks; for instance the base crucks of Glastonbury tithe barn have small open crucks rising from the collar. Often the upper roof construction rose from a second collar with the plates trapped between the two. Base crucks were in use between the thirteenth and fifteenth centuries though most are believed to be of fourteenth-century date. They are found in buildings of high status such as monastic tithe barns and manorial halls. They seem to be restricted to the Midlands

24. Aisled barn, Wycollar, Lancashire

In contrast to the aisled barns of the eastern and south-eastern counties with their steeply pitched roofs and their comparatively slender members, this building shows the low-pitched roof and the limited number of heavy timbers found in the stone-walled aisled barns of Lancashire and Yorkshire.

of England, to North Yorkshire, to Montgomeryshire and to South West England.

The hammer beam roof is the most extravagant and the least probable of the ways of achieving a long span with short pieces of timber. There are many variations but the basic hammer beam roof may be considered as a truss with principals, collar and tie-beam but with the central part of the tie omitted leaving the two ends as hammer beams. These were supported by braces rising from the walls and in turn supported hammer posts, which rose to the collar, while arch-braces ran from hammer posts to collar. In Westminster Hall, the most spectacular example of a hammer beam roof, additional arch-braces made a long ascent from the wall to the collar. (100) In other examples there were two or three sets of short hammer beams rising one above the other.

25. Church of St. Wendreda, March, Cambridgeshire
A musical angel decorates the wall piece.

26. Church of St. Wendreda, March, Cambridgeshire
This is a double hammer-beam roof except that the upper hammer beams do not support posts. Decorative wooden angels mark the ends of the hammer beams.

In yet other examples there are false hammer beams which lack hammer posts. The great majority of hammer beam trusses are found in churches and especially in Essex and East Anglia, but some are found in secular buildings. (25, 26) The hammer beam roof truss in the Banqueting Hall of Hampton Court Palace built in 1531–6 is one of the latest.

The queen post roof truss may be found in seventeenth-century buildings— exposed in utilitarian buildings, concealed in churches and assembly halls— and allowed for wider spans than those available with king post roof construction. In queen post roof construction, inclined principals rising from the ends of a long tie-beam terminated at the heads of queen posts symmetrically placed on each side of the central third or so of the roof. The heads of the posts were steadied by a stout straining beam, the principal rafters were braced by inclined braces coming back to the feet of the queen posts except that in the very widest spans there were other smaller struts, called princess posts, which rose vertically from tie-beam to principal rafter and had their own braces. The very long tie-beam of a queen post roof truss was usually of imported softwood and carpenters' guides of the eighteenth and nineteenth centuries included elaborate details for taking apart such timbers and re-assembling

them with the aid of keys, scarfing joints, bolts and fish plates.

During the eighteenth century more and more metal was used in the construction of these wide span trusses until in the nineteenth century little structural timber was used in the most elaborate trusses.

Roofing of wood and metal

The use of metal, cast iron and wrought iron, in conjunction with timber in roof construction had been known in Italy and elsewhere on the Continent during Roman times and from the late medieval period onwards. British carpenters were slow to adopt this combination of materials. One reason was the high cost of iron as a raw material, the high cost of labour in working it into a form useful to the carpenters and the unreliability especially of cast iron. Another reason was probably the craftsman's pride in trying to get the utmost out of a single material whose properties were well understood. However, the greater scarcity of native hardwoods in large sizes, coupled with increasing demands especially for large span roofs of relatively low pitch, together with a change in the relative costs of timber and iron meant a greater and greater use of the two materials combined in eighteenth-century Britain. This led to the eventual elimination of timber for wide-span roofing in the nineteenth century. Only with the discovery of new techniques, especially of glued and laminated timber in the twentieth century, has the material returned to large scale use.

The tie-beam was the most troublesome member in most roof construction. In any roof truss it was the largest member, it had to play its part in the triangulation of the truss but it also had to contend with its own weight—and the problem of sagging was very serious in large spans. One way of dealing with the problem which was popular, at least with the authors of carpenters' guides and text books, was by taking apart the proposed timber tie-beam and reconstituting it either by scarfing or flitching. In scarfing, the beam was cut more or less horizontally, many interlocking rebates were worked and the whole was reassembled with the aid of iron bolts. In flitching, the beam was cut vertically, the two halves turned about so that the heart timber was on the outside, a wrought-iron plate or plates inserted as a flitch between the two halves and the whole assembly bolted together again. Alternatively tie-beams were made of lengths of timber joined with the aid of iron fish plates bolted through the timbers.

Another impetus for introduction of iron into roof carpentry came with the gradual appreciation of what was really happening in the roof structure. This led initially to the use of iron at the joints. In king post or queen post roof construction the tendency for the post to come away from the tie-beam was countered, either through the use of long iron bolts rising from a head at the bottom of the tie-beam to a nut screwed through a slot in the post, or through the use of wrought-iron straps enveloping the tie-beam and secured by gibs and cotters through the post. The tendency for the joint at the foot of the truss to come apart as the principal rafter left the tie-beam was countered similarly either by the use of a bolt or wrought-iron straps.

The tendency for other joints to move was countered by the use of two-legged or three-legged wrought-iron plates bolted through the members meeting at the head of king post or queen post. As casting techniques improved elaborate cast-iron shoes were introduced for receiving the ends of tie-beams and principal rafters, with provision for bolting down to padstones or longitudinal beams. These shoes helped to deal with problems of shear as heavily loaded trusses met narrow walls or columns.

As the understanding of the forces acting in roof trusses increased so it was realised that composite trusses, using timber for compression members and wrought-iron for tension members, were feasible and might be economical. One candidate for early replacement was the timber king post. At least on the shorter spans the king post with all its elaborate joints and wasted timber could be replaced by a simple wrought-iron rod flattened at each end for bolting to the timbers. The other candidate was the tie-beam. Where serving as a restraining member, helping to keep the walls of the building in position, it had to be stout and heavy but where it was simply the main tension member of a roof truss then it could be replaced by a wrought-iron rod, or later by an angle section in rolled steel, and simply bolted to principal rafters and king rods. Throughout the late eighteenth and nineteenth centuries improvements in the manufacture of iron-based materials continued apace so that one by one the hand-based operations of the blacksmith were superseded by the machine-based operations of the iron or steel mill. These improvements meant better, cheaper and more versatile iron or steel components and reduced timber members to a more specialised and less significant role.

Flooring, beams and girders

The problems arising from the need to provide intermediate floor construction in multi-storey buildings rarely led to solutions with anything like the spectacular results found in much roof construction but the problems were real enough and there was some variety in their solution.

A roof had to carry the weight of rain or snow which might lodge on its surface, the weight of roof covering and its own self-weight. Except in the case of the small proportion of flat or nearly flat roofs to be found on older buildings the designer of a roof could take advantage of the roof pitch to provide triangulation. Except for the desire of the late medieval and renaissance designers to conceal the roof, virtually any pitch within the range available for a given roofing material could be used to help economise in roof construction. An intermediate floor, however, had to carry its own weight and the weight of furniture, fittings, machinery, goods etc. which might be found in the building but it might also be called upon to carry the very considerable and unpredictable weight of many people. It had to carry these loads within a very limited depth and almost without the benefit of triangulation: deep floor zones using triangulated flooring beams were virtually unknown until recently. The floor also had to play its part in the total structure of most buildings helping to prevent posts or walls from buckling vertically and preventing solid walls from collapsing outwards. In addition, many floors were expected

27. Smithills Hall, Bolton, Lancashire
The ceiling of this room reveals the construction of the intermediate floor. The floor-joists, which are set flat rather than upright, are carried by binders which in turn are jointed into a girder. The joists are splayed with simple stops and both binders and girder are heavily moulded.

to provide a decorative ceiling to be seen from the level below. In some cases it was intended that floor construction would allow for sound insulation between floor levels. All these requirements had to be met with the same eye for economy, the same need to work within the limitations of the timber available as in any other branch of carpentry.

Flooring or so-called 'naked flooring' varied in complexity mainly according to the spans involved. In the simplest floors, relatively thin floorboards were carried by floor-joists spanning from wall to wall. In floors of wider span a beam or 'binder' ran from wall to wall carrying joists and boards. In the largest floors of greatest span a very heavy beam or 'girder' ran between supports and carried a series of binders which in turn carried joists and boards. (**27**)

Normally, floorboards were about one inch (25 mm) thick and ran across the floor-joists to which they were nailed. However in early floor construction fairly narrow boards, about 6 or 7 inches wide (150 or 170 mm) were let into rebates running along the length of the joists, the floor finish thus consisting of boards and exposed upper parts of joists. Early or utilitarian floorboards were butted together along their length. Later and more sophisticated techniques led to the introduction of loose tongues of wood let between grooves along the boards and eventually to the machine-produced integral tongued and grooved boards with which we are still familiar. The ends of floorboards were normally butted together but in superior work of the late eighteenth and nineteenth centuries the ends were spliced together or linked by means of loose double-ended dovetails let into the boards. Again, in superior work there was 'secret nailing' whereby boards were nailed through tongues and nailheads did not appear on the surface. During the nineteenth century it was common in all but the cheapest work to cramp the boards together before nailing them to the joists. Until cheap softwood boarding became widely available, boards of random width were laboriously cut from the hardwood—oak or elm—and laid so that the widest boards were in the middle of the floor, gradually working out to the narrowest boards at the perimeter. In the most superior work a subfloor of conventional boarding was laid and then a decorative floor of patterned boards or blocks perhaps of more exotic timber was superimposed. Generally, however, the carpenter was most concerned that his floorboards gave a smooth upper surface. Boards simply cut as slices from a log were liable to warp as they dried out; those cut by quartering across

28. Pest House (Plague House), in churchyard at Great Chart, Kent
Here the wall consists of large panels a storey in height and a bay in width.

the grain were more satisfactory but were in narrow widths and expensive to produce as the conversion entailed much waste. Softwood timber from the Baltic was sought for the best work and well-seasoned 'Estrick Board' might be used even when all the other timber in a building was locally produced.

Floor-joists changed in shape from the flat or square section to the tall narrow section still used. The joist which was square or approximately square in section was theoretically wasteful since a tall narrow section best resists the tendency to bending under load but until machine-sawn softwood came into use the section was, in fact, a good way of using the timber available. Joists were selected from trunks or branches which could be adzed or axed to the required size and shape rather than being cut from large trunks by means of laborious and expensive sawing. Quite significantly the square-section joist was not liable to collapse from sideways buckling nor from toppling sideways. Nevertheless the deeper cross-section was more economical and eventually won the day.

For the deepest joists, however, it was necessary to take precautions to avoid buckling and sideways collapse. This was done by introducing 'strutting' or 'bridging' at mid-span. The cheapest form of strutting consisted of short lengths of boarding knocked into position between the joists. The best form of strutting was the herringbone whereby short lengths of batten were nailed in a criss-cross pattern at the midspan of the joists. These types of strutting were widely used in the nineteenth century; they may be concealed between floorboards and plaster ceiling in domestic buildings but are usually exposed in the floors of industrial or farm buildings.

In double floors the joists were given intermediate support by the binder, which was a heavy beam usually of square section and usually cut from the trunk of a tree. Its dimensions were determined by floor loading and self-weight but its shape was also affected by the method of carrying the joists. In theory the joists could run over and be carried on top of the binder and, in fact, this was done from the later nineteenth century onwards but in practice joists in short lengths were jointed into the binder. At first the housing joint employed cut into the top of the binder, which had to be correspondingly wide, but, later, haunched tenon joints were used which required the loss of less timber and that from the less important centre of the binder.

For the largest spans, and in triple floors, heavy beams called girders were used to support the binders. A girder had to carry the load of a large part of the floor, together with its own self-weight, over a considerable span and without sagging, and this even though it was liable to be weakened at a vital point by the joints connecting girder and binder. Carpenters had to search far to find suitable timber. Long straight softwood timber of great cross-section was much prized; one often sees such timber re-used and betraying in empty mortice holes and rebates the original use. Alternatively tie-beams from long discarded medieval roof trusses may be seen in cellars carrying the ground floor construction in much later buildings. Girders also provided a suitable subject for flitched and similarly strengthened timbers.

Where the floor was to be part of a new building then naturally provision could be made in the posts of a timber wall, or by way of recesses, padstones or corbels in solid walls, for carrying the binders and girders. Where large floor areas with the minimum of solid partitioning was required—as in a textile mill or a maltings—then timber posts or cast-iron columns were included in the design so as to cut down unsupported spans and, making use of timber pillars or cast-iron shoes, to master the joints. However, sometimes floors had to be inserted into existing buildings; the practice of introducing an intermediate floor into a medieval hall was common, for instance, and the structurally separate samson post shows how this could be done.

Floors often had openings as for staircases or fireplaces and then the floor construction had to be 'trimmed', joists being cut short as 'trimmed joists' and let into a 'trimmer joist' which was itself carried by a 'trimming joist'. In the best quality work trimmer and trimming joists were of stouter cross-section than other joists but in ordinary domestic construction a uniform joist section was used as far as possible.

The underside of a floor gave an opportunity for display, either by decoration on the timber itself or by way of a plastered ceiling applied to the timber. Medieval carpenters had expected users of their buildings to look upwards in the great halls and admire the dimly lit intricacies of exposed roof timbers. In rooms with intermediate floors there was an even greater opportunity for elaborate decoration; the timbers were lower and could be better lit. So from medieval times and right through the sixteenth and seventeenth centuries all important rooms had decorated ceilings and most of these included decoration of the structural timbers. Joists were chamfered with simple stops, binders had more elaborate mouldings based on the chamfer with appropriately complicated stops. Some binders and girders when used had successions of mouldings based on the roll moulding to give very complicated exposed surfaces. The chamfer and its derivatives were appropriate on all types of timber because they acknowledged the problem of the waney edge as a consequence of squaring the logs. Where binder and girder intersected sometimes the mouldings were interrupted, using the 'mason's mitre' technique so that one member acknowledged the other. Alternatively, decorated bosses (carved blocks of timber) were applied to mask the joint. Sometimes non-structural timbers were introduced to divide the ceiling into diamonds or octagonal coffers. Sometimes (though more often in pattern books than in actual buildings) binders were cut short to support each other in a sort of maze.

The exposed structural ceiling was largely superseded by the plaster ceiling. At its simplest this still left main members exposed, but more commonly a decorative pattern ran right across the ceiling. The plaster was carried on ceiling joists which were quite separate from the floor-joists though carried on the same binders. In the late eighteenth and nineteenth centuries, small rooms with little or no ceiling decoration nevertheless had separate ceiling joists; they minimised cracking from floor loads and helped to reduce sound transmission from one room to another. The weight of the floor construction and the correspondingly greater degree of sound insulation was increased by the use of 'pugging', a layer of sand carried on boards between the floor-joists.

Partitions

Within the building, the carpenter was responsible for the timber partitions which helped to define the various rooms and other spaces. These partitions could be supported by other members, be self-supporting, or could be structural partitions helping to support some other part of the structure.

One type of partition common in medieval buildings and in seventeenth-century farmhouses was the 'muntin and plank' partition. It consisted of a row of vertically set planks, alternate ones being thick and grooved to receive the remainder which were thinner and resulted in a rhythm of thick and thin, narrow and wide panels which could in turn be enlivened by mouldings or paintwork. In a rather cheaper version, the thin panels were replaced by wattle and daub giving a construction and effect similar to that of narrow-panelled half-timbering. The muntin and plank partition rose from a timber sill at floor level to a timber beam spanning from wall to wall which in early examples existed simply to receive the planks, but in most cases also served as a binder carrying floor joists.

Another type of partition used from the medieval period right into the nineteenth century was the timber-framed partition. Here the construction familiar in half-timbered buildings was employed within the building. This was sensible enough in timber-framed buildings, but the technique was common in brick or stone buildings also. Construction was of studs and rails making narrow or square panels as in walls; the wattle and daub work ran between the panels in half-timbered partitions but in the eighteenth and nineteenth centuries it was more common to fill in the panels with brick and run plaster on laths over the whole surface of the partition. Such partitions might be self-supporting at intervals within the building or part of a structural cross-frame, helping to brace the main walls and form part of the roof structure.

The third type of partition, the trussed partition, was in vogue in the late eighteenth and nineteenth century and is still in occasional use today. The trussed partition was in effect a very deep beam composed of small section timbers and carrying a load, including its own weight, over a gap. A trussed partition, thus, would have been used where a series of smaller rooms on an upper floor was to be located over an open room on a lower floor. Instead of carrying live and dead loads on a deep beam which might adversely affect the ceiling design or even the proportions of the room, they were carried through the triangulation of the partition. The design also allowed for openings in the partition: very large openings were difficult to accommodate but door openings could be accommodated in the triangulation. Carpenter's guides and building text books of the nineteenth century abounded in ever more ingenious designs for trussed partitions; their construction was made easier as iron straps and bolts became cheap and widely available. Trussed partitions survive in many buildings of the period as a trap for the unwary of the present day who alter buildings without fully understanding their history or construction.

Later developments

The direct line of development from the heavy timber-framed construction of the medieval period into the eighteenth and nineteenth centuries was one in which heavy members assisted, and in some cases replaced, by iron components were used to produce quite complex constructions solving reasonably large building problems. However there had always been some element of lightweight construction as in mud and stud buildings and, although no direct connection is claimed, there were vigorous developments in lightweight timber-frame construction during the eighteenth and nineteenth centuries which have continued to the present day.

Development was based on the so-called 'uniform scantling' principle and occurred to some extent in this country but mostly in North America. The carpenters sought economies through using sawn timber of a single cross-section (usually 4 inches by 2 inches, 100 mm by 50 mm) and the simplest possible joints. Timbers were doubled or tripled where extra strength was needed, joints were lapped or halved rather than cut as mortice and tenon, and dovetail joints were avoided; nails were used to supplement or replace cut joints. A great impetus was given through the large-scale building programme in North America. In the seventeenth and early eighteenth centuries, construction there had been based on English or European precedent and this tradition continued, though in a diminishing proportion of buildings. At the same time horizontal log construction was used both for temporary and permanent buildings in those extensive parts of the continent in which the right sort of timber was available. But the aim was to have a framed building and this came to mean a uniform scantling frame of softwood timber with nailed joints and of materials transported from source to site with the help of railways and waterways.

In the parts of North America east of the Great Plains, and in much of the West Coast, softwood timber was abundant and water power for sawmills was generally plentiful. The manufacture of wire cut nails made the other main component cheap and plentiful. Two variations of the uniform scantling technique were devised and continue in use: balloon framing and platform framing. In both cases the frame consisted of 4 inch by 2 inch studs (100 mm by 50 mm) rising from a 4 inch by 2 inch sill, doubled at the corners and around openings, given triangulation by diagonally set boarding known as sheathing nailed to the outside of the studs, given weather protection by horizontal boarding nailed to the sheathing, given draught protection by plaster or matchboard lining and given fire protection by 'fire stopping' of 4 inch by 2 inch rails at intervals to break up the cavities between the studs. In balloon framing the studs rise through two storeys (or more in a multi-storey building) and intermediate floor-joists were supported partly on a timber ribbon let into the studs and partly by face-nailing between joist and stud. In platform framing the studs rose only through one floor, ending in a plate which received the floor-joists and on which rested another plate serving as a sill to the next storey of studs.

Roof construction was also based on the single cross-sectional dimension and the use of nails. Among the finest achievements of uniform scantling construction in North America must rank the circular or polygonal barns, rising several storeys and depending for their wall and roof construction on the stability which comes from their shape and the use of wire cut nails to join together lengths of 4 inch by 2 inch softwood timber.

Uniform scantling construction was used in cottages and other small buildings in the south-eastern counties of England during the nineteenth century but to nothing like the extent of use in North America. Brick was fashionable, cheap and almost universally available; softwood timber was an imported material, requiring regular maintenance and of doubtful fire-resistance. Recently there has been a revival of interest in uniform scantling timber-framed house construction of what is essentially platform framing, but with a brick skin giving a durable and acceptable outward appearance. The main advantages of this composite construction are speed of erection, economy of labour on site and easy provision of heat insulation between the studs.

Roof construction employing uniform scantling softwood timber has also developed over the past forty or fifty years quite apart from these developments in walling techniques. During the first half of the twentieth century it had become normal, especially in house construction, to carry rafters on purlins supported by brick partition walls. This was satisfactory and economical but meant that house planning was limited by the need to provide partition walls where they would suit the roof construction, which was not always where they would suit the house plan. A simple form of roof truss was devised by the Timber Development Association in the 1940s which freed the plan since all roof loads were carried by the outer walls. The roof truss in fact depended on a sort of trussed rafter occurring at every fourth rafter interval and supporting a pair of light purlins, which in turn gave intermediate support and a measure of longitudinal restraint to the common rafters. The size of members in trussed rafters and common rafters was reduced to 3 inches by 2 inches (75 mm by 50 mm) because of the scarcity of timber and in the trussed rafters the members were placed face to face and secured by means of bolts and split-ring connectors. Although scientifically designed, the TDA roof (as it was generally called) was in fact related to the traditional roof construction of Lowland England with its clasped purlins, trussed rafters and use of truss members as common rafters.

Another form of roof construction was developed in North America and introduced to this country in the 1960s; it is now in universal use in house construction and is also related to the roof construction of lowland England. This is generally known as 'gang-nail construction' from one of its chief versions. Trussed rafters are made up of light timber sections and the joints are either made with the aid of plywood gussets, carefully nailed according to a calculated pattern or, more commonly, with the aid of thin galvanised steel plates cut to produce a scientifically-designed pattern of spikes hammered into the timber. Such trussed rafters are carefully but economically manufactured in factories, and, being light in weight, are easily transported and handled on the building site. They are analogous to traditional trussed rafter roofs and can suffer from the same problems of longitudinal restraint.

Although not often seen in house construction these various ways of using timber of uniform scantling, put together with nails, plates, bolts or connectors may be seen in other building types: in certain industrial buildings, in buildings where an element of prefabrication is desirable and feasible and especially in farm buildings where timber construction suits the function, the location, the degree of accessibility and the available labour.

One further development in timber construction which is worth mention, though it is quite different from uniform scantling, is that in which timber is broken up and then reconstituted in a more useful form. The most familiar instance is plywood whereby thin layers of timber peeled from a log are glued together with opposing grain to produce a cheap and durable sheet material. Blockboard is similar but with short lengths of small timbers glued between plywood faces. In chipboards the fibres of timber are taken apart and reconstituted as a thin slab which can replace boarding for floorboards and sheathing, being more uniform, more stable and quicker to lay on site. On a larger scale, beams or complete frames may be made by glueing together pieces of timber: purlins or floor binders, for instance, may be made by glueing together timber and plywood in a box section; complete beams or half-trusses may be made from laminations of timber glued together. In fact a revival of cruck construction may be seen in the use of laminated timber frames.

Conclusion to Part One

It is tempting to consider timber-frame construction as a single entity: there is a single material—timber—and it is being considered in use for a single purpose—the enclosure of space by building. But it is clear that in considering timber-frame constructions even in Britain alone we have to look at several developments. Some proceed simultaneously with others, possibly by way of separate schools of carpentry, each with its origin in a distant and long-forgotten past, or possibly by way of a single set of techniques widely known and adopted by the carpenters according to their assessment of the circumstances of each case. We have to accept in this study as in so many others the opposing pulls of tradition and innovation, tradition suggesting the well-understood, the well-proven solution to a familiar problem, innovation suggesting a new, untried, possibly more economical solution to a problem which may not be so familiar after all. We have also to accept that carpentry was never static. The availability of timber, the raw material, varied constantly; there were competing users even for structural quality timber; the supply of home-grown timber seemed to be diminishing in spite of tree-farming and efforts towards general conservation; the supply of imported timber was for long uncertain, available only at certain ports and at uncertain cost. The demand for the carpenter's time varied with the building cycles, the long cycles lasting decades or even centuries, the shorter cycle of a few years boom and a few years decline, the yearly cycle of active and slack building seasons. The effects of general technological progress are not really understood: how much of the development of jointing techniques depended on improved quality of tools; how much of the development from pegged to bolted joints depended on a demand for cheap bolts on the part of the carpenters or a supply of cheap bolts from the blacksmiths who found they could apply steam power to machinery invented originally for other purposes. We take for granted that roofs will be supported by timber roof trusses, yet the roofs of barns and domestic buildings were often supported by walls or buttresses or arches located for that purpose. We take for granted half-timbered wall construction as the obvious and economical way of using the available material but we are only gradually coming to appreciate the significance of mud and stud construction, the poor relation of what might well have originated as an elite constructional technique.

Yet certain general trends seem to be obvious enough. There was indeed a development from earth-fast timber construction to one involving framing and with self-sustaining structures independent of the ground on which they sat. Not all the phases in this development have been ascertained but stages at each end have been studied and excavation and speculation will presumably fill the gaps. There was indeed a phenomenon which we recognise as cruck construction: it seems to have materialised fully developed, remained in maturity for several centuries and then quite swiftly declined, disappeared or changed into other carpentry techniques. Efforts to provide antecedents have so far proved unconvincing; modern analogies are coincidental rather than part of a continuous development.

Another general trend, not of course confined to timber-frame construction was for a substantial but diminishing element of traditional experience— pragmatic, sometimes drawing incorrect conclusions from misleading evidence—to exist alongside a limited but increasing element of imported or scientifically determined knowledge. The obvious instance is the king post perpetuated as a compression member while already being regarded as a tension member. Not all carpenters were backward, hidebound by tradition, just as not a few were innovators, constantly trying to advance the craft.

We also have to beware of the tendency, again not confined to the study of timber-frame construction, for the classification and typology prepared with the benefit of hindsight to have been equally clear to those working at some time in the past. For instance, one is tempted to assume that in highland areas of Britain cruck construction was superseded by the more complicated but more versatile box-frame and post-and-truss constructions whereas it is clear that often the two techniques were used side by side on the same site, sometimes in the same building, apparently with function, beauty or utility in mind. Thus the post-and-truss house in Cheshire might be served by a contemporary cruck-trussed barn or the single farm building might have three unobstructed bays of cruck construction at the barn end with a couple of bays of hay loft over cow house composed of posts, tie-beams, principals and floor beams at the other.

A further complication in the study of timber building lies in the limited range of dating criteria. Some few timber buildings or timber members are dated, other pieces of timber can be dated with varying degrees of accuracy by dendrochronology or radiocarbon techniques. Other timber-framed buildings or buildings with timber roofs can be convincingly dated from documentary evidence, but the vast majority of timber structures can only be approximately dated if at all. The pioneering work on dating by jointing or mouldings (where present) must be much further developed before we can

29. Weobley, Herefordshire
Among the timber-framed buildings in the picture may be seen one with an exposed cruck blade, the bottom part cut off (house in front of the church); a building with jettied first floor showing a cranked tie-beam and sharply curved braces; simple square-framed panelling; and, on the extreme right a brick building painted to imitate timber-framing but possibly concealing a timber-frame.

date fragments of ostensibly sixteenth-century timber with the same accuracy which archaeologists seem to be able to apply to fragments of Roman pottery.

All these complications, however, add to the interest of the study. It would not be unprecedented for the solution of one problem to lead to further unsuspected problems. But one can only trust that further understanding, deeper study, wider appreciation and much more preservation of timber building will ensure that this study is indeed continued and enhanced. (**28**) (**29**)

Part Two:
Illustrated Glossary
of Terms and Techniques

The glossary consists of terms used for items and techniques in timber construction including jointing and roof construction. The terms are mostly those used in England but some used in Wales, Scotland and North America have been included. Some of the terms are traditional, others are those which have been coined (principally by R. A. Cordingley, C. A. Hewett or J. T. Smith) to deal with items for which no traditional term is known or for which traditional terminology is misleading. Except where the meaning is self-evident a sketch or photograph has been included though many photographs illustrate several terms.

Although every effort has been made to ensure consistency, sometimes the accepted use has been inconsistent. This is true especially of the terms 'post' and 'strut'. Generally a post carries a longitudinal member such as a wall-plate and a strut a lateral member such as the principal rafter in a roof truss, but this is not always so. Queen posts, for instance, usually carry purlins but not invariably, while Princess posts are really struts helping to carry roof truss members.

An attempt has been made to bring the material up to date by including a selection of items from contemporary timber construction but most of the techniques and terms are those of traditional carpentry.

Where there are several terms for the same item the selected term is given first and the others follow in brackets.

ABUTMENT: the point or surface at which the end of one piece of timber touches another. A butt joint is one in which two ends meet. (d10)

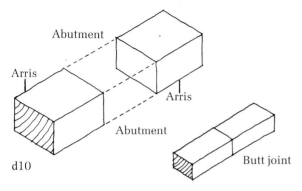

d10

AISLE: in certain timber-framed buildings a range running along one or both sides, or ends, of a space open to the roof. The aisles are covered by continuations of the main roof. An aisle differs from an outshut in that the space enclosed flows freely between the main space and the aisle, whereas an outshut is walled or partitioned from the main space. (d11)

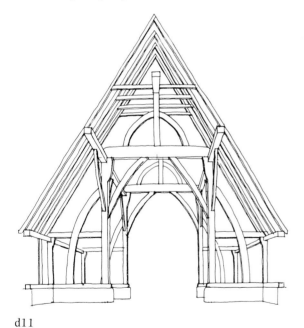

d11

AISLED BUILDING: a building with aisles along one or both sides and possibly one or both ends in addition. Normally a row of posts separates the main space from the aisles. (d11, d12) (30) (31)

FULLY AISLED buildings have aisles on both sides. (d12)

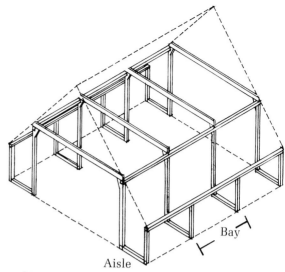

d12

SEMI-AISLED buildings have aisles on one side only. (d13)

QUASI-AISLED buildings have no row of posts between the central space and the aisles. (d14)

QUASI-SEMI-AISLED BUILDINGS lack posts between the main space and its single aisled extension. (d15)

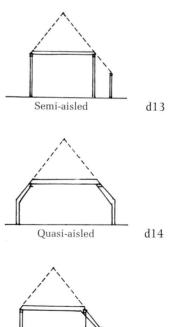

Semi-aisled   d13

Quasi-aisled   d14

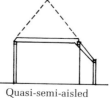

Quasi-semi-aisled   d15

AISLE PLATE: a member laid horizontally along the top of an aisle wall and receiving the feet of rafters which form part of the roof of the aisle. (d16a)

AISLE POST: *see* ARCADE POST *under* POST

AISLE TRUSS: a roof truss carried on arcade posts in an aisled building

ANCHOR BEAM: *see* BEAM

ANGLE POST: *see* POST

ANGLE STRUT: a short inclined member intended to act under compression while reducing the unsupported length of a major structural member. (d17)

30. Aisled barn, Matching Hall, Matching, Essex

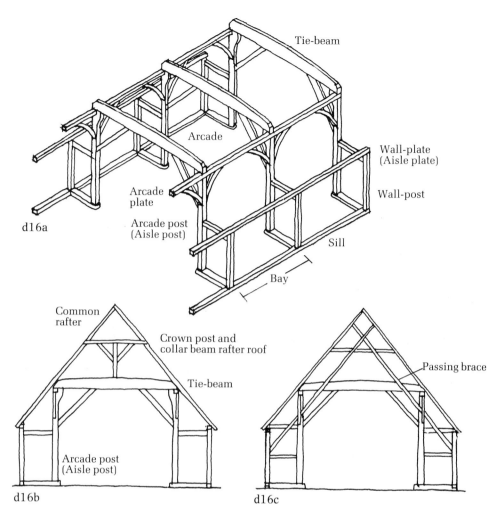

Tie-beam

Arcade

Arcade
plate

Wall-plate
(Aisle plate)

Wall-post

d16a

Arcade post
(Aisle post)

Sill

Bay

Common
rafter

Crown post and
collar beam rafter roof

Tie-beam

Passing brace

Arcade post
(Aisle post)

d16b

d16c

ANGLE TIE (ANGLE BRACE, DIAGONAL BRACE, DIAGONAL TIE, DRAGON TIE): a piece of timber placed across the angle at the corner of a hipped-roof building, tying together the wall-plates and receiving the inner end of a dragon beam. (d18c)

ARCADE: the range of posts running between the main span and the aisles of an aisled building and so braced to the arcade plate as to give the effect of a series of arches. (d16a)

ARCADE PLATE (ROOF PLATE): a member set square (i.e. horizontally and not at right angles to the roof slope) and running along a line of arcade posts in an aisled building, so giving intermediate support to common rafters. The arcade plate corresponds to the wall plate in a building without aisles. (d16a,b)

ARCADE POST: *see* POST

31. Aisled barn, Gawthorpe Hall, Lancs.

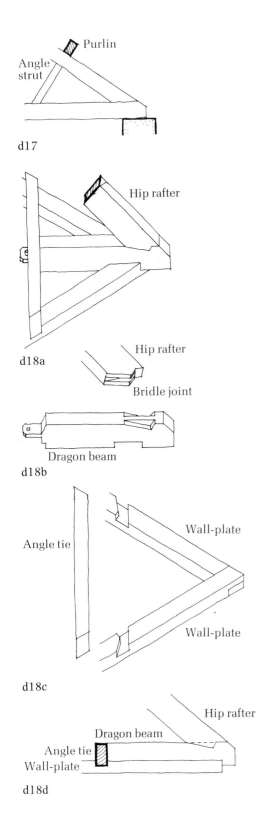

Purlin

Angle
strut

d17

Hip rafter

d18a

Hip rafter

Bridle joint

Dragon beam

d18b

Wall-plate

Angle tie

Wall-plate

d18c

Hip rafter

Dragon beam

Angle tie

Wall-plate

d18d

ARCH-BRACED ROOF TRUSS: a truss consisting essentially of a pair of principal rafters and a collar with arch braces running between principal rafters and collar. The arch braces may rise from, at, or near the wall-plate level or may rise from lower down the wall. (d19a,b) (**32**)

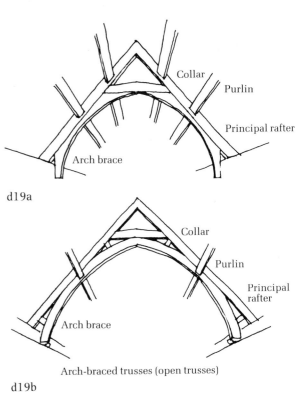

Collar

Purlin

Principal rafter

Arch brace

d19a

Collar

Purlin

Principal
rafter

Arch brace

Arch-braced trusses (open trusses)

d19b

ARRIS: a sharp angle formed at the meeting of two surfaces of a piece of timber. (d10)

ASHLARING: the use of short vertical timbers rising from the inner surface of a wall to meet the common rafters. Ashlaring helps to conceal what would otherwise be an unsightly gap between wall head and rafters. Sometimes the ashlaring carries boarding or lath and plaster. In attics or garrets the ashlaring provides low side walls between floor and rafters. (d20) (34)

ASHLAR PIECE: a short stud rising from a sole piece or an inner wall plate to a common rafter as part of a run of ashlaring. (d20)

ASHLAR POST: a short post running along the line of the inner face of a solid wall and rising to join the underside of a principal rafter. (d21)

32. Arch-braced roof truss, Lytes Cary, Somerset.

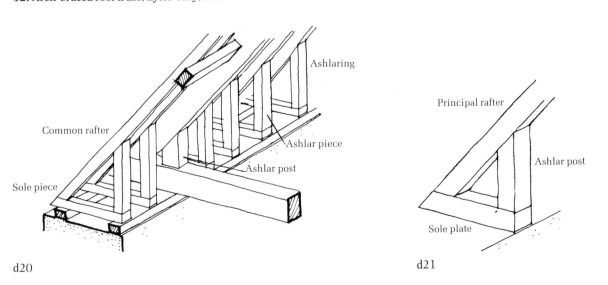

d20

Ashlaring

Common rafter

Ashlar piece

Ashlar post

Sole piece

d21

Principal rafter

Ashlar post

Sole plate

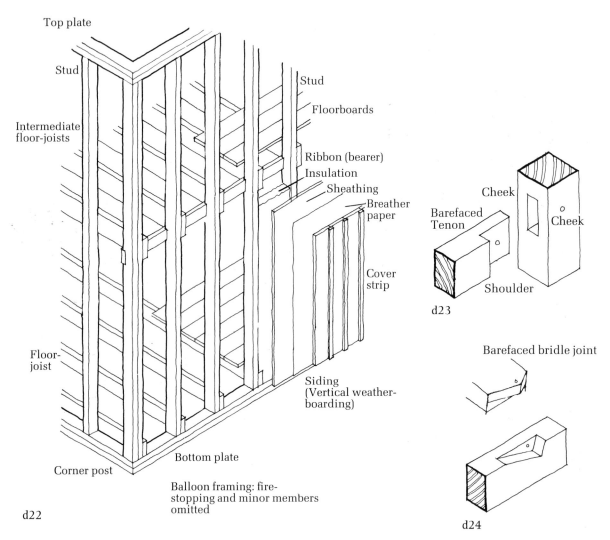

Top plate

Stud

Intermediate
floor-joists

Stud

Floorboards

Ribbon (bearer)

Insulation

Sheathing

Breather
paper

Cover
strip

Floor-
joist

Siding
(Vertical weather-
boarding)

Corner post

Bottom plate

Balloon framing: fire-
stopping and minor members
omitted

d22

Cheek

Barefaced
Tenon

Cheek

Shoulder

d23

Barefaced bridle joint

d24

ASSEMBLY MARKS: *see* CARPENTERS' MARKS

BALK (BAUK, BAULK, BAWK): the squared trunk of a tree
intended for use as a beam, usually between about 8 ins
and 12 ins square (203 mm to 305 mm). Such a piece
of timber might be used as a tie-beam which in Scotland
and the North of England was traditionally called a 'balk'.
During the eighteenth century the term was used for large
squared tree trunks shipped or floated from the Baltic.

BALLOON FRAMING: a type of lightweight timber-frame
construction developed in North America in which sawn
studs, usually 4 ins by 2 ins (100 mm by 50 mm) run
from a timber sill through two or more storeys to finish
at a wall-plate. First floor joists are carried on a light timber ribbon let into the studs and are face-nailed into the
studs. Triangulation is given through the use of diagonal

boarding or plywood sheathing nailed to the outsides of
the studs. The construction depends on nailing rather
than jointing.

The technique is believed to have originated in Chicago
in the 1830s. The term balloon framing was a con-
temptuous reference to an apparently weak and insub-
stantial form of construction which, in fact, was strong
and robust. (d22) *see also* PLATFORM FRAMING

BAREFACED: refers to a piece of timber in a joint which displays one shoulder instead of the normal two. The term
is most commonly applied to a barefaced tenon where one
shoulder rather than two butts against the cheeks of the
mortice. (d23, 24) *see also* JOINTS

BARGE BOARD (VERGE BOARD): a board which follows the
incline of a gable end, protecting otherwise exposed tim-

33. Barge boards, Porch House, Potterne, Wilts.

bers, terminating the roof covering and giving an opportunity for decorative treatment. (**33**)

BASE CRUCKS: pairs of naturally curved timbers rising from, at, or near ground level and with the curved parts inclining inwards to terminate at a heavy collar. Base crucks may be considered as truncated crucks (as the term suggests) or as posts with inward-inclining upper parts. (d25, 26a)

> BASE CRUCK TRUSS: in a base cruck truss the collar carries square-set purlins which may be locked into position by means of a second collar. There may be straight or curved braces between base cruck blades and collar. Above the base cruck collar various methods of roof construction may be used, including a rafter roof with crown post and collar purlin and a small cruck truss and trenched purlins 'riding' on the collar. (d25) (**34**)

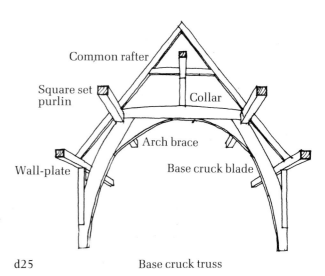

d25　　　　　　　　　　　　Base cruck truss

95

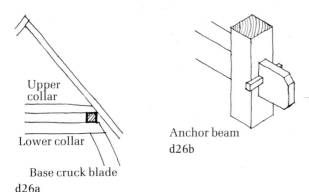

Upper
collar

Lower collar

Base cruck blade
d26a

Anchor beam
d26b

BATTEN: a piece of timber of rectangular cross-section, usually about 1 in (25 mm) thick and between 2 ins (50 mm) and 4 ins (100 mm) wide. A batten is bigger than a lath and smaller than a plank.

BAY: the unit of division applied to a wall frame, a roof frame or both together. In a wall frame a bay is the space between two sets of principal posts. In a roof frame a bay is the space between two trusses. Thus a building may be described as of three bays if there are two pairs of intermediate posts between the end walls, and a roof may be considered to be of three bays if there are two trusses between the gables. (d12, 65)

The width of a bay may vary considerably over the country as a whole but was often fairly uniform in a given building-type in a particular place at one period, and so houses and barns are often described in documents as being 'of so many bays'.

The term HALF-BAY is sometimes used to describe a structural division which is much narrower than those in the remainder of a building.

BEAM: the term is usually taken to refer to a heavy piece of timber tying together two parts of a structure and/or carrying a superimposed load, in either case while laid horizontally.

ANCHOR BEAM: a beam whose end passes as a tenon through a post and is pegged. (d26b)

BOX BEAM: a type of built-up beam consisting of top and bottom sections of solid timber glued to plywood sides leaving a hollow between. (d27a)

BUILT-UP BEAM: a beam composed of two or more pieces of timber secured together to give greater strength than either would have separately. As used in the eighteenth and nineteenth centuries the technique was akin to scarfing and involved the insertion of keys, or serrating or indenting adjacent surfaces, to prevent sideways sliding of the pieces of timber; in addition bolts or hoops were used to keep them securely connec-

ted. More recently, built-up beams have been composed of several pieces of timber screwed, bolted or glued together. (d27b, c)

CAMBERED BEAM: a beam in which one or more surfaces are curved upwards towards the centre. CAMBER may refer to the slight curve of the underside of a beam to counteract any appearance of sagging, or it may refer to the curve of the upper part of a beam thickening the centre and appearing to increase its strength (a shape sometimes called HOGBACK). It may also refer to the curve of both upper surface and underside of a beam intended to give the effect and appearance of an arch. (d28a, b, c)

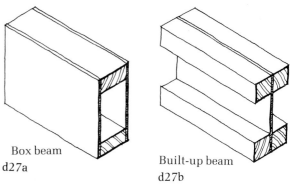

Box beam
d27a

Built-up beam
d27b

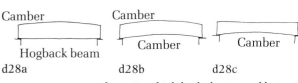

Camber

Hogback beam
d28a

Camber

Camber

d28b

Camber

Camber

d28c

CRANKED BEAM: a beam in which both the top and bottom surfaces are angled towards the centre so giving a cranked appearance to the whole beam. (d29)

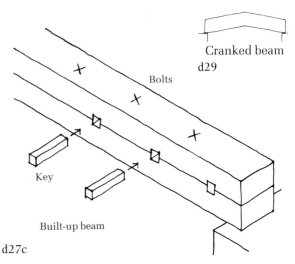

Cranked beam
d29

Bolts

Key

Built-up beam
d27c

DRAGON BEAM (q.v.): a beam set diagonally at the corner of a building.

FLITCHED BEAM: a type of beam made by taking a balk of timber, sawing it in half, turning the two halves back to back to reveal the heartwood, inserting a steel or

34. Base crucks, tithe barn, Barton Farm, Bradford on Avon, Wilts.

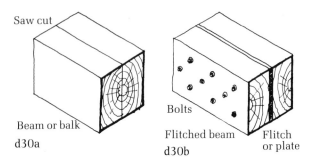

Saw cut

Beam or balk

d30a

Bolts

Flitched beam

d30b

Flitch or plate

wrought-iron plate between the two pieces and then bolting the three together. Such a beam was stronger than the original balk. (d30 a, b)

GIRDING BEAM: an alternative term for GIRTH (q.v.).

HAMMER BEAM (q.v.): receives a PRINCIPAL RAFTER at one end and a HAMMER POST at the other.

MANTLE BEAM: a light beam sometimes found between the blades of an OPEN CRUCK TRUSS.

SILL-BEAM: *see* SILL

TIE-BEAM: a beam tying together the post-heads of a timber-framed wall or the upper surfaces of a solid wall. The tie-beam may also form part of a roof construction as when supporting a crown post or receiving the ends of blades or principal rafters. (d31a, b)

TRUSSED BEAM: a deep beam of composite construction usually with the upper part of timber, acting in compression, and the lower part of a wrought-iron rod acting in tension and kept in position by a strut or struts made either of timber or cast iron. (d32)

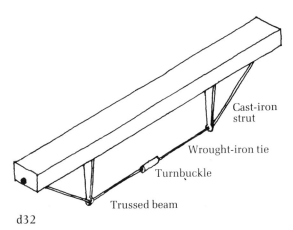

Cast-iron strut

Wrought-iron tie

Turnbuckle

Trussed beam

d32

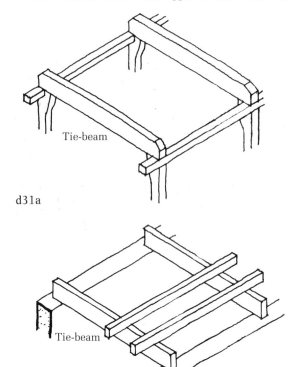

Tie-beam

d31a

Tie-beam

d31b

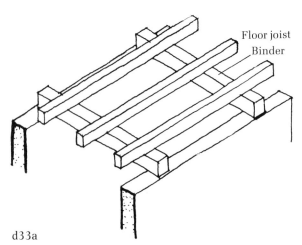

Floor joist

Binder

d33a

BEARER: a term used mostly in North America for a light-weight beam secured to double-height studs and receiving first floor joists. (d22)

BELFAST ROOF TRUSS: a lightweight roof truss having a segmental arch shape to the upper chord and a flat lower chord and a set of battens placed criss-cross in between. Such trusses were popular in roofing small industrial buildings in the late nineteenth and early twentieth centuries.

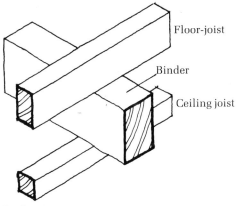

d33b

BINDER (BINDING BEAM, BINDING JOIST): a heavy member running horizontally from post to post or wall to wall and giving intermediate support to floor-joists and, sometimes, ceiling joists. In some text books a binder is called a BRIDGING JOIST though in others that term is used for common joists. (d33a, b)

BIRDSMOUTH (SALLY): a notch cut from the end of an inclined timber, such as a common rafter, as it bears on a horizontal timber such as a wall-plate. There is some resemblance to the open beak of a bird. (d34)

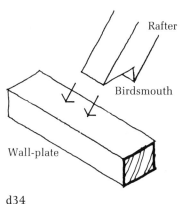

d34

BLADE (BACK, TRUSS BLADE, RAKING BEAM): a term applied to the heavy inclined member of a roof truss believed to have been developed from, or related to cruck construction. Purlins are usually trenched into the upper surfaces of blades. It is the equivalent of a principal rafter in trusses derived from other forms of construction. *See also* CRUCK BLADE *under* CRUCKS

BLOCKBOARD: *see* PLYWOOD

BLOCKING PIECE: a piece of timber laid solidly upon the back of a blade to make up the difference between the intended pitch of the roof and the actual angle of inclination of the blade. Such timbers are most often used as part of cruck construction (q.v. *under* CRUCKS).

BOND TIMBERS (CHAIN TIMBERS): lengths of timber, usually about 3 or 4 ins deep (76 or 100 mm) laid in a solid wall to provide horizontal reinforcement. Such timbers are widely used in countries subject to earthquakes but were also used in Britain to help avoid the cracking which resulted from settlement of walls or foundations. Bond timbers might also serve as battens for fixing decorative linings to a building but that was not their main purpose. *See also* GROUNDS

BOOM: the top or bottom horizontal member of a TRUSSED BEAM or TRUSSED PURLIN.

BOSS: an ornamental block carved out of the solid timber or applied at the junctions of timbers in a roof, ceiling or floor. Bosses range in date from those used in medieval roofs (including wooden vaulting) to those used in decorative timber ceilings of the sixteenth and seventeenth centuries. (d35) (**35**)

BOX BEAM: *see* BEAM

35. Boss, ceiling at Daisybanks, Shibden, Yorks. W.R.

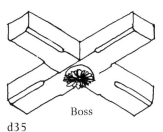

d35

BOX-FRAME CONSTRUCTION: a term often used nowadays to refer to a form of construction in which the walls of a building are framed out of horizontal and vertical timbers (usually with the aid of diagonals) to produce a wooden box. The roof acts as a lid to the box but may or may not be structurally related to each of its parts. Where the box is developed from pairs of timber posts at the two ends and at intermediate points along the length of the building, and where these posts mark the positions of roof trusses, then the term POST-AND-TRUSS is now preferred. (d36a, b)

BRACE: a subsidiary member in a timber frame located at an angle between two main members and stiffening them by triangulation. A brace may also act as a strut in reducing the unsupported span of a horizontal load-bearing member. (d37a) (37)

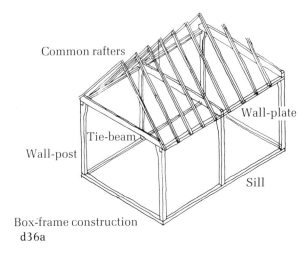

Box-frame construction
d36a

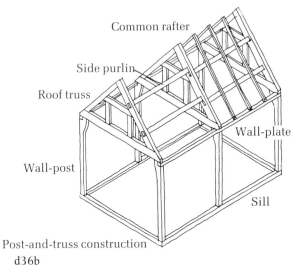

Post-and-truss construction
d36b

36. Arch brace, gatehouse at Bolton Percy, Yorks. W.R.

ARCH BRACE: a brace running between a vertical and a horizontal member but with the lower surface, at least, curved in an arch shape. In roof construction an arch brace may rise from a wall or post to a tie-beam or collar above. In wall construction an arch-brace may rise between a post and a wall-plate. In roof construction the arch brace is often solid between principal and collar but in wall construction the arch brace usually leaves a gap between posts and wall-plate. (d19, 25) (36)

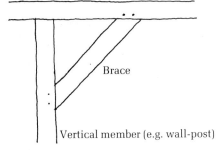

Horizontal member (e.g. wall-plate)

Brace

Vertical member (e.g. wall-post)

d37a

FOOT BRACE: a term sometimes used for a down brace between post and sill.

KNEE BRACE: a type of short up brace rising from a post to a wall-plate, or, in roof construction, from a wall-post across a tie-beam to meet a principal rafter. (39)

OGEE BRACE: a brace with a double curve. A NORMAL OGEE meets a horizontal member approximately at right angles; a REVERSED OGEE leaves a vertical member approximately at right angles. (d37g)

PASSING BRACE: the term coined to describe a light-weight member which runs diagonally across several main horizontal and vertical members and serves to

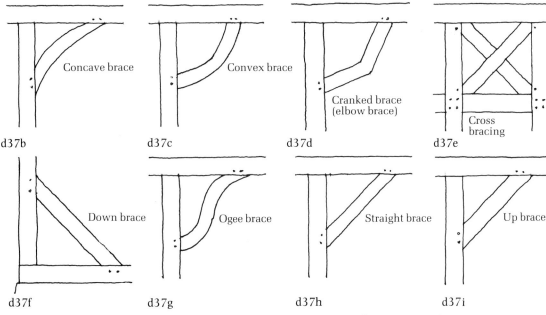

Concave brace

d37b

Convex brace

d37c

Cranked brace (elbow brace)

d37d

Cross bracing

d37e

Down brace

d37f

Ogee brace

d37g

Straight brace

d37h

Up brace

d37i

CONCAVE BRACE: a brace curved into the angle which is being crossed. (37b)

CONCEALED BRACING: is halved to the insides of studs and concealed from outside view, (41c)

CONVEX BRACE: a brace curved out from the angle being crossed. (d37c) (40)

CRANKED BRACE (ELBOW BRACE): a brace with a sharp natural bend producing a blunt point within the angle being crossed. (d37d)

CROSS-BRACING: the use of braces in two directions to form a St. Andrew's cross. (d37e)

DOWN BRACE: a brace which falls from a vertical member (such as a post) to a horizontal member (such as a sill). (d37f) (45)

ELBOW BRACE: see CRANKED BRACE

restrain them against sideways movement. A passing brace may be simply nailed or pegged to the surface but is usually housed into the main members. (d16c) (d38) (43)

PURLIN BRACE: see WIND BRACE

SCISSOR BRACE: used in pairs, usually in a rafter roof, the scissor braces are lightweight members which cross diagonally between pairs of common rafters and serve both to triangulate and to reduce the unsupported length of the rafters. (d39)

SLING BRACE: a diagonal member running from a post up to a principal rafter and receiving the end of a truncated tie-beam. (d40)

STRAIGHT BRACE: a brace running in a straight line between the members to be braced together. (d37h)

SWAY BRACE: see WIND BRACE

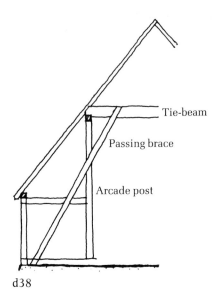

Tie-beam

Passing brace

Arcade post

d38

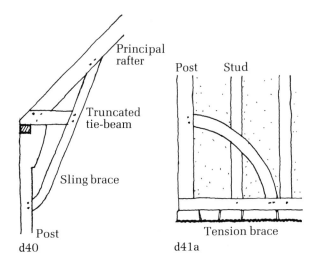

Principal rafter

Post   Stud

Truncated tie-beam

Sling brace

Post

d40

Tension brace

d41a

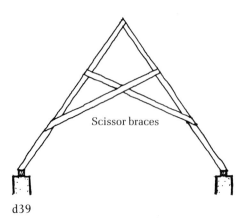

Scissor braces

d39

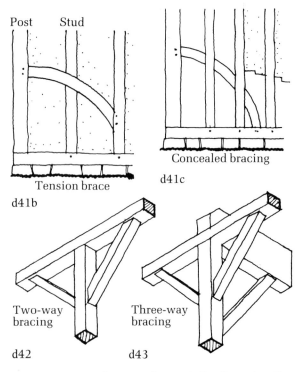

Post   Stud

Tension brace

d41b

Concealed bracing

d41c

Two-way bracing

Three-way bracing

d42   d43

TENSION BRACE: a convex brace falling from a post to a sill or from a post to a stud. (d41a,b) (**37**)

THREE-WAY BRACING: bracing which extends from an upright in three directions. Bracing from an arcade post to arcade plates on each side and to a tie-beam would provide an example of three-way bracing. (d43) (**31**)

TWO-WAY BRACING: bracing which extends from an upright in two directions. Knee braces rising from each side of a wall-post would be an example of two-way bracing. (d42) (**38**)

UP BRACE: a brace which rises from a vertical member (such as a post) to a horizontal member (such as a wall-plate) or from an inclined member (such as a principal rafter or blade) to a horizontal member (such as a purlin). (d37i) (**41**)

WALL BRACE: a brace used as part of wall construction rather than roof construction.

WIND BRACE (PURLIN BRACE, SWAY BRACE): a member which embraces the angle between a blade or principal rafter and a purlin. A wind brace is usually an up brace but may be a down brace. (d44a,b) (**42**)

BRACKET: a short piece of timber projecting from one mem-

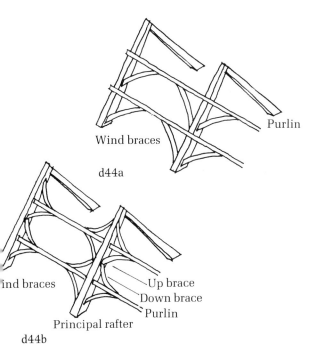

Wind braces

Purlin

d44a

ind braces

Up brace
Down brace
Purlin

Principal rafter

d44b

ber to support another. A bracket may be cut out of the solid but is usually jointed into the member from which it projects. A CONSOLE BRACKET has volutes at top and bottom as in a classical console. (d45a, b) (46)

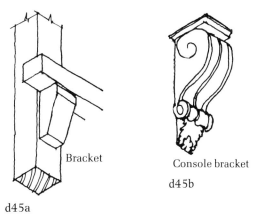

Bracket

d45a

Console bracket

d45b

37. Tension braces, Priest's House, Stratford St. Mary, Suffolk.

BREATHER PAPER: stout paper membrane water-resistant but pervious to water vapour used between sheathing and siding in timber stud walls. (d22, 240)

BRESSUMMER (BREST-SUMMER, SUMMER, SUMMER BEAM, SUMMER TREE): this term has been used with various meanings in Britain and North America.

1. A beam intended to support a wall over an opening though possibly carrying joists also; used in the face or breast of a wall which is also part of a chimney breast.

2. A horizontal timber running the length of a building or a wing of a building and carrying an upper part such as a jettied upper storey.

3. A beam within a building carrying the binders or girders of a framed floor. The term is also sometimes used for the binder which carries the floor-joists.

4. The term SUMMER has been loosely used for any sort of heavy beam in a building.

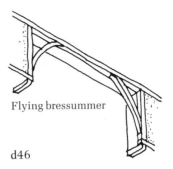

Flying bressummer

d46

38. Braces, house at Carlton Husthwaite, Yorks. N.R.

39. Braces, house at Sutton Courtenay, Berks.

40. Braces, building in West Stockwell Street, Colchester, Essex.

41. Braces, Capons Farm, Cowfold, Sussex

42. Windbraces, solar roof, Fiddleford Mill, Dorset

43. Passing brace, Grange Farm Barn, Coggeshall, Essex

FLYING BRESSUMMER: a term occasionally used for a horizontal timber carrying a deeply projecting eaves as in a Wealden House, though FLYING WALL-PLATE would seem to be a more appropriate term for this detail. (d46) (44)

BRICK NOGGING: *see* NOGS

BUILT-UP BEAM: *see* BEAM

BUTT-PURLIN: *see under* PURLIN

CABER: *see* COMMON RAFTER *under* RAFTERS

CAMBER: *see* CAMBERED BEAM *under* BEAM

CANT: the facet on the underside of a roof whose angles are cut off by collar, soulaces and ashlar pieces. (d47)

CARBON 14 DATING: *see* RADIOCARBON DATING.

CARCASS: a timber frame as left by the carpenter and before finishing by the joiner, plasterer etc.

 CARCASS FLOORING (NAKED FLOORING): the frame of timber assembled in place ready to receive floorboards and possibly a separate ceiling.

 CARCASS ROOFING: the frame of roof timbers assembled

44. Flying bressummer, Monks Barn, Newport, Essex

45. Braces, Old School House, Bunbury, Cheshire

107

46. Bracket, Trewern, Montgoms.

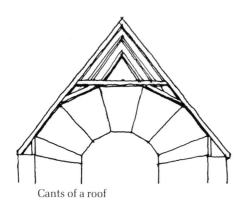

Cants of a roof

d47

in place and ready for the laths and slates or other roof covering.

CARPENTER (WRIGHT, HOUSE-WRIGHT): a craftsman skilled in the shaping and framing together of timber for building purposes. Since WRIGHT was a general term for all craftsmen who worked in wood (including cartwright, wheelwright, shipwright, millwright etc.) the term was often used to describe a carpenter. A carpenter working in structural timber should be distinguished from a JOINER (q.v.) working in non-structural wood.

CARPENTERS' MARKS (ASSEMBLY MARKS): these are symbols scratched, incised or chiselled into timber usually to assist in assembly and re-assembly but occasionally to indicate that timber has been worked by a particular craftsman. (d48a, b, c, d) **(47) (48)**

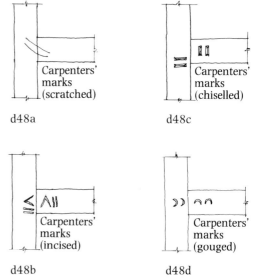

d48a

d48c

d48b

d48d

CARPENTRY: the theory and practice of preparing and assembling timber in constructional frames for buildings.

CATSLIDE ROOF: *see* ROOF SHAPES

CHAMFER (CHAMPHER): the splayed face resulting from the removal of the angle or arris along a piece of timber. (d49a)

> HOLLOW CHAMFER: a splay cut as a concave rather than as a flat surface. (d49b)

> OVOLO CHAMFER: a splay cut as a part-round rather than as a flat surface. (d49c)

> SUNK CHAMFER: a splay recessed from the plane surfaces of the timber. (d49d)

47. Carpenters' marks, gatehouse at Bolton Percy, Yorks. W.R.

48. Carpenters' marks, Melville House, Fife, Scotland

CHAMFER STOP: the detail at the end of a chamfered stretch marking the return to a square section.

BAR STOP: the termination of a chamfer with the aid of a transverse half-roll. (d50a)

BROACH STOP: the termination of a chamfer with the aid of a half-pyramid. (d50b)

FLAT STOP: termination of a chamfer in a flat triangle. (d50c)

JEWELLED STOP: facetted decoration projecting from the end of a chamfer. (d50f)

LAMB'S TONGUE STOP: the termination of a chamfer by a doubly curved surface narrowing to an arris. (d50d)

LEAF STOP: termination of a chamfer in a leaf shape. (d50e)

NICKED STOP: the use of an additional cut beyond the main stop in an arris. (d50h)

OGEE STOP: termination of a chamfer by two curves:

Flat chamfer
d49a

Hollow chamfer
d49b

Ovolo chamfer
d49c

Sunk chamfer
d49d

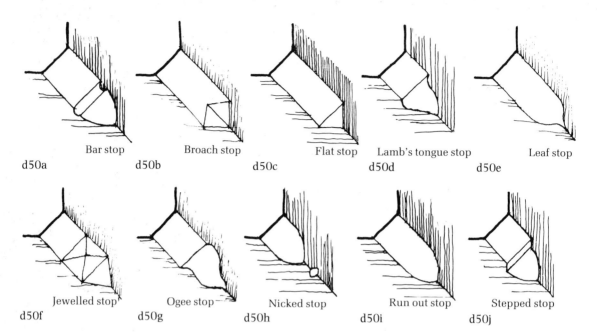

Bar stop	Broach stop	Flat stop	Lamb's tongue stop	Leaf stop
d50a	d50b	d50c	d50d	d50e

Jewelled stop	Ogee stop	Nicked stop	Run out stop	Stepped stop
d50f	d50g	d50h	d50i	d50j

convex and then concave. *cf.* LAMB'S TONGUE STOP. (d50g)

PYRAMID STOP: *see* BROACH STOP.

RUNOUT STOP: the chamfered surface dies out gradually in a slight curve to become an arris. (d50i)

STEPPED STOP: a chamfer changing to a fresh plane before the stop proper. (d50j)

CHASE: a long groove cut into a piece of timber. (d151)

CHEEK: the solid timber left on each side of a mortice hole and receiving the shoulders of a tenon. *see also* JOINTS and JOINTING. (d23)

CLADDING: the covering sometimes applied to the outer surface of a timber frame. Thus weather-boarding, for instance, may be nailed to the posts and studs of a wall to provide a weather-board cladding.

MATHEMATICAL TILE CLADDING: a type of cladding in which specially shaped tiles were applied to timber-framed walls. The tiles were moulded in such a way as to allow overlapping but with the exposed face shaped like the face of a brick (a full brick, a half-brick or a closer). They were nailed to laths or boards but usually given a plaster bedding. Pointed joints helped to convey a convincing imitation of a brick wall. (d51a) (**49**)

PLASTER CLADDING: a type of cladding in which a rendering and finish of lime plaster was applied to a

timber-framed wall. Usually the rendering was made on to split oak laths nailed to the studs but in early and inferior work hazel withies were used. Often plaster cladding was applied to existing timber-framed walls, but the technique was popular especially in the eastern counties of England from the early seventeenth century onwards in new construction as well as in refurbishment of older buildings. The technique of 'pargetting' i.e. making incised or moulded patterns in the plaster is a decorative variation of plaster cladding. (d51b) (**50**)

SLATE HANGING: a type of cladding in which slates were hung on laths nailed to a timber-framed wall. The slates were sometimes squared roofing slates but often slates cut to a fish-scale or serrated shape were used. (d51d) (**52**)

TILE HANGING: a type of cladding in which plain tiles were hung on laths nailed to a timber-framed wall. Usually the plain tiles were ordinary roofing tiles, but often tiles specially shaped to give a scalloped or serrated effect were introduced as an addition to plain tiles or in substitution for them. (d51c) (**13a**)

WEATHER-BOARD CLADDING: weather-boarding may be let into panels or used as cladding laid horizontally or vertically. (**51**)

49. Mathematical tile cladding, building in West Street, Faversham, Kent

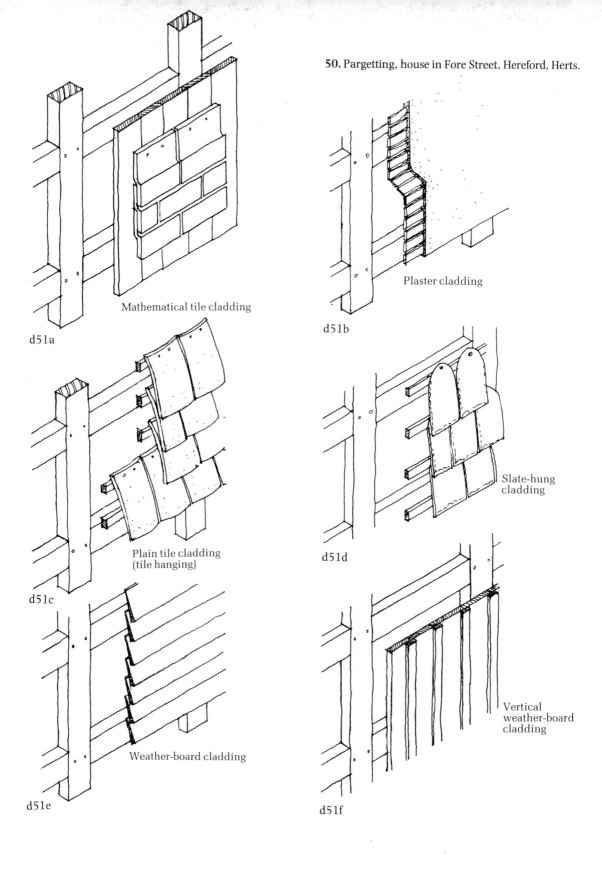

50. Pargetting, house in Fore Street, Hereford, Herts.

Mathematical tile cladding

d51a

Plaster cladding

d51b

Plain tile cladding
(tile hanging)

d51c

Slate-hung
cladding

d51d

Weather-board cladding

d51e

Vertical
weather-board
cladding

d51f

112

51. Weather-board cladding, house at Smarden, Kent

52. Slate cladding, The Nunnery, Dunster, Somerset

1. Weather-boarding may consist of butt-edged boards fitted into grooves in the sides of studs and forming the panel infill to half-timbered walls.

2. Weather-board cladding may consist of boards, usually feather-edged, nailed on to studs in a timber-framed wall so as to overlap each other.

3. Boards, usually butt-edged, may be nailed vertically to the rails of a timber-framed wall and, with the vertical joint protected by a cover strip, may serve as a cladding. (d51e,f)

> BEADED EDGED BOARDING consists of boards, not tapering in cross-section, which are grooved near the bottom to form a continuous bead.

> CLAP BOARDING: the term used in North America for weather-boarding. Clap boards were used by coopers in making barrels and the term may have

been adopted when similar boards were used to form a continuous weatherproof and draughtproof surface to timber-frame buildings.

FEATHER-EDGED BOARDING: Boards with a tapering cross-section nailed horizontally to studs with the thick bottom part of a board overlapping the thin upper part of the board beneath.

SIDING: An alternative term used in North America for weather-boarding.

CLAPBOARD: *see* WEATHER-BOARD CLADDING *under* CLADDING

CLASPED PURLIN: *see* PURLIN

CLEAT: a block of wood attached to one member and locating or fixing another. The most common application is to the topside of a blade or principal rafter to keep a purlin in place. Cleats may be housed into the supporting members but often are simply spiked or bolted into position. (d52)

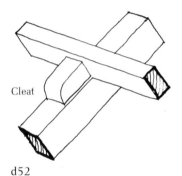

d52

CLOSE-COUPLE RAFTER ROOF: *see* RAFTER ROOF

CLOSED CRUCK TRUSS: *see* CRUCK TRUSS *under* CRUCKS

CLOSE-STUDDING: *see* PANELS

CLOSE-STUDDING WITH MIDDLE RAIL: *see* PANELS

COLLAR (COLLAR BEAM, COLLAR PIECE, COLLAR RAFTER, STRUT BEAM, STRUTTING BEAM, WIND BEAM): a horizontal member in a roof spanning between a pair of inclined members—blades, principal rafters or common rafters—and located about halfway between wall-plate level and apex. In some circumstances a collar is a tie or tension member helping to prevent the rafters from spreading apart, in others a collar is a stiffening or compression member helping to prevent the rafters from sagging inwards.

COLLAR BEAM: this term suggests the heavier collar used as part of a roof truss. (d53)

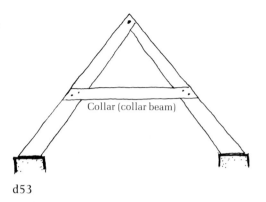

d53

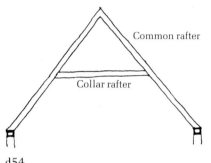

d54

COLLAR RAFTER: this term is used where the collar is of the same section as the common rafters and is used as part of a system of trussed rafters. (d54)

COLLAR AND TIE-BEAM TRUSS: a roof truss consisting of a pair of inclined blades or principal rafters rising from a tie-beam and including a collar (or occasionally two collars) which may receive various vertical or inclined struts. (d55)

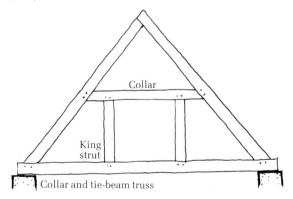

d55

COLLAR PURLIN (COLLAR PLATE) *see* PURLIN

COLLAR RAFTER ROOF: *see* RAFTER ROOF

COMMON PURLINS (HORIZONTAL RAFTERS): *see* PURLINS

COMPOSITE ROOF TRUSS: a roof truss comprising timber members and also members of metal, usually wrought iron, sometimes cast iron and occasionally of steel. A roof truss consisting of principal rafters and diagonal struts of timber and a king rod and tie rod of wrought iron, would be an example of a composite roof truss. (d56)

COUPLE-CLOSE ROOF: *see* RAFTER ROOF

COUPLE TRUSS: a roof truss consisting only of pairs of principal rafters at bay intervals. The couples serve also as common rafters and may be diminished to common rafter size above a purlin. (d57)

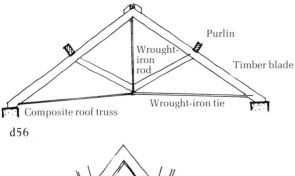

d56

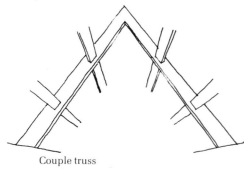

Couple truss

d57

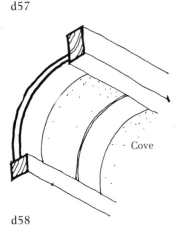

d58

COUPLED RAFTER ROOF: *see* RAFTER ROOFS

COVE: a curved underside to an overhang, as under a jetty. (d58) (**53**)

CROSS-BRACING: *see* BRACE

CROSS-NOGGING: *see* STRUTTING

CROWN POST: a post standing centrally on a tie-beam to support a collar purlin. The post may be of a simple square cross-section or shaped as in an octagonal or a cruciform shape, for instance. It may be plain from top to bottom or have a carved shoulder and base. There may be a jowl at one or both sides to receive the collar purlin. (**54**)

The crown post may be free-standing but is usually braced by down braces running from post to tie-beam. Usually there are up braces between crown post and collar purlin but there may also be bracing from the crown post to a collar giving four-way bracing at the head of the crown post.

The term crown post was formerly used as an alternative to king post (q.v.), a member running between tie-beam and ridge, but nowadays the term is restricted to the shorter post which ends at the collar level. (d59, 60, 61, 62, 63, 64)

CROWN POST AND COLLAR PURLIN ROOF: *see* RAFTER ROOFS

CROWN POST RAFTER ROOF: *see* RAFTER ROOF

53. Cove, Bents Cottage, Mottram St. Andrew, Cheshire

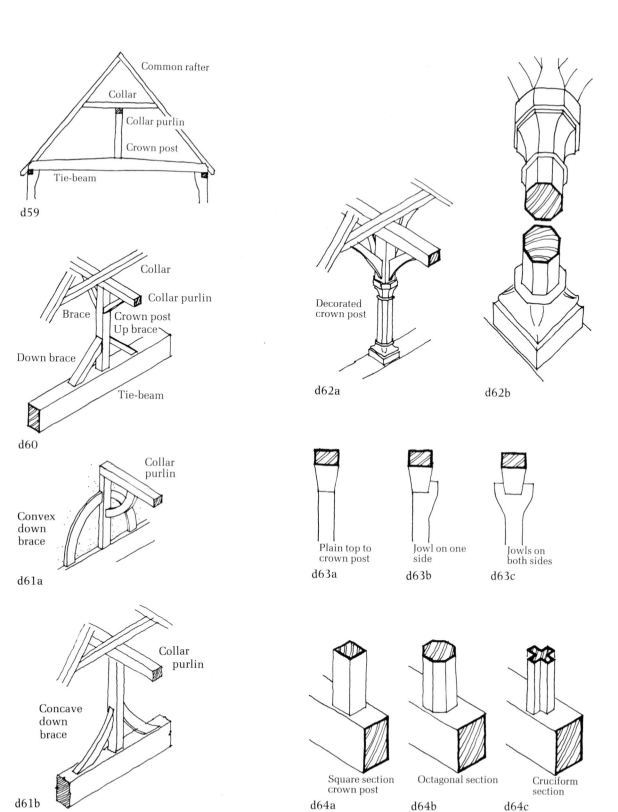

Common rafter
Collar
Collar purlin
Crown post
Tie-beam

d59

Collar
Collar purlin
Brace
Crown post
Up brace
Down brace
Tie-beam

d60

Collar purlin
Convex down brace

d61a

Collar purlin
Concave down brace

d61b

Decorated crown post

d62a

d62b

Plain top to crown post
d63a

Jowl on one side
d63b

Jowls on both sides
d63c

Square section crown post
d64a

Octagonal section
d64b

Cruciform section
d64c

CRUCKS (CROOKS, CROCKS, CRUTCHES, FORKS, SILES): pairs of timbers, usually heavy and of large cross-section, rising from, or near, ground level to meet at, or near, the apex of a roof. Each individual cruck is called a blade (or, in the North of England, a sile) and may be straight, elbowed, double or single-curved, and is usually tapered. Pairs of blades were often cut from the same tree, sawn and turned to make a matching pair. Alternatively two separate timbers were roughly matched. (d65) (**55**)

CRUCK BLADE: one of a pair of crucks.

END CRUCK (HIP CRUCK, HIP POST): where a hipped roof

54. Crown post, Durham House, Great Barfield, Essex

was desired in a cruck-trussed building a half-cruck was introduced. This consisted of a single curved or elbowed blade rising to meet the end of a ridge purlin. (d66) (**56**)

EXTENDED CRUCKS: crucks whose blades have been extended by means of a length of timber jointed above the collar to reach the apex while maintaining a true cruck profile. (d67)

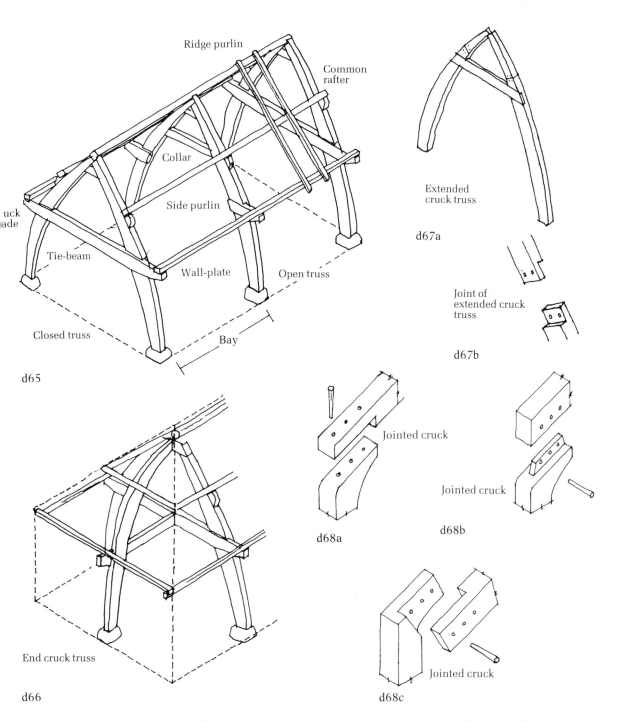

Ridge purlin

Common rafter

Collar

Side purlin

uck ade

Tie-beam

Wall-plate

Open truss

Closed truss

Bay

d65

Extended cruck truss

d67a

Joint of extended cruck truss

d67b

End cruck truss

d66

Jointed cruck

d68a

Jointed cruck

d68b

Jointed cruck

d68c

JOINTED CRUCKS: cruck blades made of more than one piece of timber and with the joint well below collar level. Usually one part resembles a post with its end tilted inwards while the other part resembles the blade of a roof truss. Jointed crucks of the highest quality have a tenon rising from the post half of the jointed cruck and engaging in a mortice in the upper half, while side pegs run through the mortice and tenon joint. This type of jointed cruck is sometimes called a scarfed cruck in that the joint resembles a scarfed joint

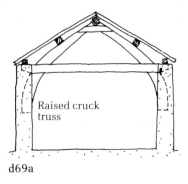

Raised cruck truss

d69a

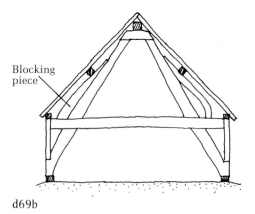

Blocking piece

d69b

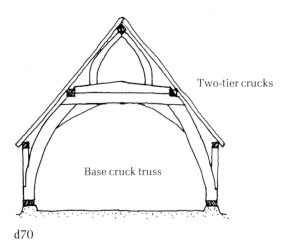

Two-tier crucks

Base cruck truss

d70

55. Cruck truss, Dairy Farm, Mobberley, Cheshire

56. End cruck (hip cruck), barn at Corrimony, Inverness-shire, Scotland

57. Jointed crucks, barn at Preston Plucknett, Somerset

120

(q.v.). Jointed crucks of lesser quality have the upper portion carried on the inclined part of the lower portion and the two are face-pegged together. In the crudest version the two portions are simply lapped side by side and have a few side pegs to hold the parts together. (d68) (**57**)

RAISED CRUCKS: cruck blades rising from part way up a solid wall i.e. usually from 5 ft (1524 mm) or more above ground level. (d69a) (**58**)

SCARFED CRUCKS: a term sometimes used generally for jointed crucks but more properly reserved for those of superior quality which have a scarfed joint.

TRUNCATED CRUCKS: crucks ending at collar level and used in the gables of buildings with half-hipped roofs. (d71) (**60**)

TWO-TIER CRUCKS: the use of a small pair of crucks rising from the collar of a base cruck truss to complete the roof truss. (d70) (**59**)

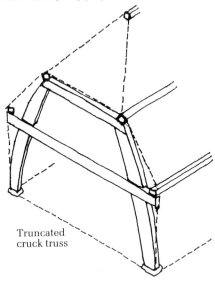

Truncated
cruck truss

d71

CRUCK SPUR: this term has been used with several meanings.
1. A short piece of timber cantilevered from a cruck blade to carry a wall-plate.
2. A short piece of timber running between a cruck blade and the post of a timber-framed wall, usually lap-jointed to the blade and mortice and tenoned to the post. Here the cruck spur is intended to help keep the timber-framed wall in position relative to the crucks. This sort of cruck spur is sometimes called a CRUCK TIE.
3. A short piece of wrought iron nailed or screwed to a cruck blade, twisted and nailed or screwed to a wall plate so keeping it in position. (d72e)

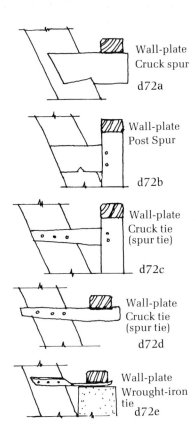

Wall-plate
Cruck spur
d72a

Wall-plate
Post Spur
d72b

Wall-plate
Cruck tie
(spur tie)
d72c

Wall-plate
Cruck tie
(spur tie)
d72d

Wall-plate
Wrought-iron
tie
d72e

CRUCK TRUSS: a pair of inclined cruck blades linked by a tie-beam, a collar or both, and rising from, or above, ground level to terminate at or just below ridge level constitutes a cruck truss. Normally there are several such trusses placed at bay intervals along the length of a building. (d65)

Almost invariably the cruck truss carried a ridge purlin and the roof covering was carried by way of side purlins and wall-plates supported by or linked to the members of the truss. Purlins may be carried on the backs of the cruck blades with a trenched joint or with the aid of the cleat, but often the side purlins were carried on the backs of blocking pieces or outer blades introduced to make up the difference between the angle of the upper part of the cruck blade and the pitch of the roof. (d69b)

CLOSED CRUCK TRUSS (FULL CRUCK TRUSS): a cruck truss which incorporates a tie-beam. (d65)

OPEN CRUCK TRUSS: a cruck truss which omits the tie-beam. Usually there is a heavy collar beam instead. (d65)

58. Raised crucks, tithe barn at Barton Farm, Bradford on Avon, Wilts.

RAISED CRUCK TRUSS: a cruck truss which rises from part way up a solid wall (i.e. where the feet are more than about 5 ft (1524 mm) above ground level). (d69a)

UPPER CRUCK TRUSS: a cruck truss which rises from a beam in an upper part of the building and into which the ends or feet of the crucks are jointed. (d73)

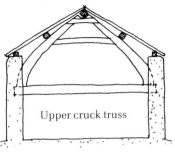

Upper cruck truss

d73

59. Two-tier crucks, tithe barn at Barton Farm, Bradford on Avon, Wilts.

60. Truncated crucks, cottage at Styal, Cheshire

CUSPS (CUSPING): a cusp is the point between foils in Gothic tracery. The foils have concave sides so forming the point of the cusp. In carpentry, cusping was often used as decoration to exposed timber both externally in decorative panelling and internally in roof construction. The members of a roof truss or the upper part of a cruck truss visible from an important room were often carved to display foliated shapes with cusps. Similarly the wind braces between side purlins were often carved to display cusps between foils. (d74) (**61**)

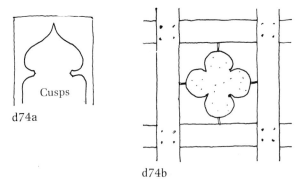

d74a

d74b

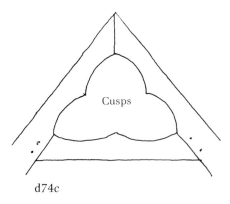

d74c

DEAL, DEALS: deal is a general term given to softwood timber used in buildings. More specifically deals are planks of softwood about 9 ins (228 mm) wide and up to 4 ins (100 mm) thick.

DENDROCHRONOLOGY: the technique of dating timber with the aid of tree rings. The technique is based on the understanding that growing timber increases in size through the addition year by year of concentric rings of timber cells, and the observation that the character and especially the width of the various tree rings depends on the climatic conditions which existed at the time and place of the growth of a particular tree. Making use of recently felled timber and other timbers felled at known dates it is possible to construct a pattern of wide and narrow rings.

Any other piece of timber once growing in the same locality can have its own pattern of tree rings compared to some portion of the standard pattern and its felling date thereby calculated.

DORMANT (DORMANT-TREE, DORMENT): a medieval term for a beam, remaining in use into the eighteenth century. The term was used for any horizontal load-bearing timber such as a purlin or bressummer. In its translation of SLEEPER the term remains in use for the timber beams which carry the loads of railway lines and as sleeper walls for the plate which gives intermediate support to ground floor-joists.

DOUBLE HAMMER BEAM TRUSS: *see* HAMMER BEAM TRUSS

DOVETAIL JOINT: *see* JOINTS

DRAGON BEAM (DRAGGING BEAM, DRAGON PIECE):
1. a horizontal piece of timber bisecting the angle formed by two wall-plates and running between the corner and an angle tie to receive a hip rafter.
2. a horizontal piece of timber running across wall-plates to support the intersection of two sills in a jettied floor and back to receive the shortened lengths of joists in the jettied floor.

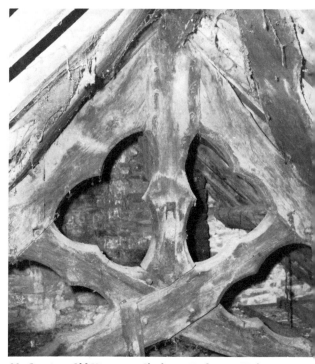

61. Cusping, Old Vicarage, Glasbury, Radnors., Wales

126

Sometimes the two terms were distinguished in that the term DRAGON PIECE was used in connection with roof construction and DRAGON BEAM in floor construction. (d18b, d)

DRAWBORING: the practice of drilling peg-holes so that they are not quite in alignment. When the peg was driven home as in a mortice and tenon joint the tenon was drawn tightly into the mortice hole. Usually the peg was deformed when driven in to draw the two parts of the joint together.

When the carpenter was trying his joints before final assembly he used a withdrawable iron drawbore pin to tighten the joint temporarily. (d75)

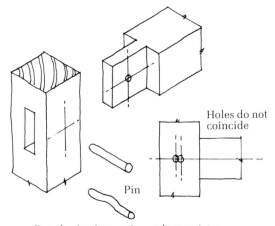

Holes do not coincide

Pin

Drawboring in mortice and tenon joint

d75

DWANG: *see* STRUTTING

EAVES: the junction between the inclined plane of a roof and the vertical plane of a wall. Usually the roof projects over the wall to a greater or lesser extent. The ends of the rafters may be displayed at the eaves or alternatively the rafters may terminate at the wall-plate and the eaves be formed by means of sprockets (q.v.). The ends of rafters or sprockets may be covered at the eaves by a plain or decorated fascia board and the projecting underside of rafters or sprockets may be concealed by a flat or inclined soffit board. Where rainwater gutters are provided they are usually fixed to the fascia board, whose depth is to some extent governed by the slope of gutter required between rainwater pipes. (d76, 77)

ELBOWED CRUCK BLADES: *see* CRUCKS

ESTRICK BOARD (ESTRICHE BOARD, EASTLAND BOARD): the name once given to softwood boarding made out of timber from Baltic countries—though the term is supposed to have been derived from the word 'Austrian'.

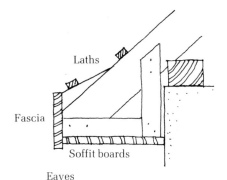

Laths

Fascia

Soffit boards

Eaves

d76

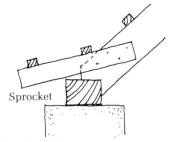

Sprocket

Sprocketted eaves
d77

EXTENDED CRUCKS: *see* CRUCKS

FACE-NAILED: in a nailed joint the use of nails driven horizontally through the vertical faces of members in order to secure them together. (d79)

FACE-PEGGED: the use of pegs in a joint in such a way that they are driven horizontally through the vertical faces of the members being jointed together. (d78)

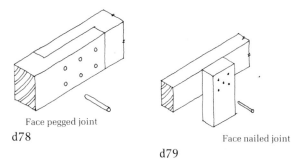

Face pegged joint

d78

Face nailed joint

d79

FALSE HAMMER BEAM: *see* HAMMER BEAM

FASCIA (FACIA): the board, plain or decorated, concealing the ends of rafters at the eaves or the ends of joists in a jettied floor. (d76, d111)

127

FASTENINGS (FIXINGS): devices other than carpentry joints used to connect one piece of timber to another.

BOLTS: wrought-iron rods forged to a head at one end and threaded to receive nuts at the other end were in use from the late seventeenth century to an increasing extent to draw together and secure, with the aid of washers, two or more pieces of timber. Bolts were also used in connection with other pieces of metal such as flitches, fish-plates and cast-iron shoes. (d80a, b)

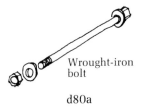

Wrought-iron
bolt

d80a

CONNECTORS: devices intended to fix securely two pieces of timber face-to-face with the aid of bolts. Toothplate, split ring and shear plate connectors are the three main types now in use. (d81, 82)

COTTERS: tapering wedge-shaped keys used together with gibs to make a tight strapped joint. (d83) (**62**)

CRAMPS (DOGS): pieces of wrought-iron bent at the ends and intended to be driven into two pieces of timber butted against each other. The ends are pointed in such a way that when the cramps are driven into two abutted timbers they are drawn together. (d84)

62. Fastenings, Herbert Warehouse, Gloucester

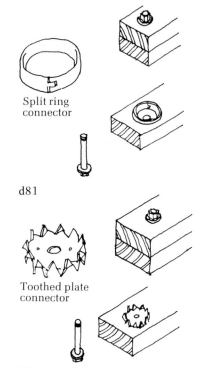

Split ring
connector

d81

Toothed plate
connector

d82

FORELOCK BOLTS: wrought-iron bolts each with a head and washer at one end and, at the other, in place of a nut and screw-thread, a slot and iron wedge for tightening. (d85)

GANGNAIL PLATE: a piece of thin galvanised steel cut in such a way that a calculated and pre-determined pattern of spikes may be driven into sets of timbers butted together and so secured. (d86a)

GIBS: U-shaped pieces of wrought iron used in conjunction with cotters to make a tight joint at a strap. (d83) (**62**)

GUSSET: a plate of wood or metal linking two or more numbers by gang nailing, glueing or nailing. (d86a, b, c)

NAILS: pointed sections of cast iron, wrought iron or steel wire of up to about 5 ins (127 mm) in length used to secure one piece of board, plank or other timber in small scantling to another.

CAST NAILS: small nails of cast iron, cheap but clumsy and brittle, once used for rough purposes such as lathing.

CUT NAILS: sheared or machine-cut or stamped from plates or rods of rolled iron or (later) steel; cut nails superseded cast nails.

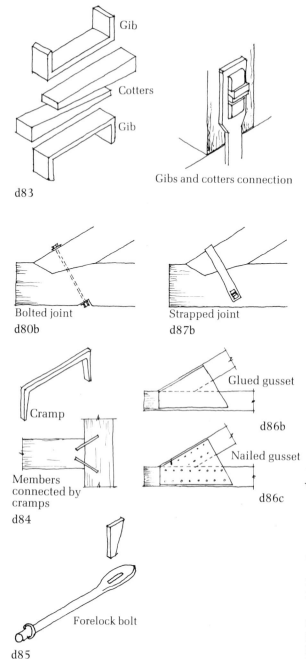

Gib

Cotters

Gib

Gibs and cotters connection

d83

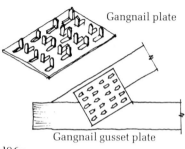

Gangnail plate

Gangnail gusset plate

d86a

to the surfaces of pieces of timber and used with the aid of bolts to secure them together. (d87c)

SPIKES: the term given to large nails over about 5 ins (127 mm) in length.

STIRRUPS: a type of STRAP consisting of metal bands, usually of wrought iron, intended to bring together pieces of timber and secured by means of gibs and cotters. (d87a, b) (**62**)

Bolted joint
d80b

Strapped joint
d87b

Cramp

Members connected by cramps
d84

Glued gusset

d86b

Nailed gusset

d86c

Forelock bolt

d85

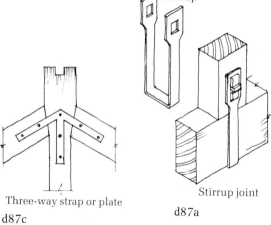

Three-way strap or plate
d87c

Stirrup

Stirrup joint
d87a

FEATHER-EDGED BOARDING: *see* WEATHER-BOARD CLADDING *under* CLADDING

FIRRINGS (FURRINGS): slips of wood nailed to the upper surfaces of rafters or joists either to bring them to a common level or to give them a uniformly sloping upper surface. Thus firrings may be used to renew a sagging floor or to provide a slope to help drain a roof. (d88, 89)

FISHED JOINT: *see under* JOINTS

FISH PIECE (FISH PLATE): a piece of wood or metal bolted or otherwise fixed to the face of timbers to be connected end to end. (d118)

WIRE-CUT NAILS: these are cut from steel wire then pointed at one end and cold-forged into a head at the other end.

WROUGHT NAILS: these are forged from narrow square rods of iron.

PLATES: specially shaped pieces of wrought iron applied

129

Firring

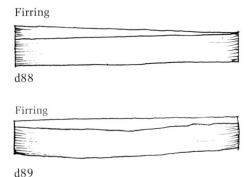

d88

Firring

d89

FISH PLATE JOINT: *see under* JOINTS

FLAT ROOF CONSTRUCTION: roofs which were absolutely flat
were almost never constructed. Instead roof finishes were
laid sloping to ensure proper drainage. The normal construction comprised joists laid horizontally, firring pieces
added to provide the desired slope and then boarding on
to which the lead or other roof covering was applied.

FLEAKS: rough boards used in scaffolding.

FLITCHED BEAMS: *see* BEAMS

FLOORBOARDS: thin timber planks laid on joists to provide
a floor surface. Early floorboards were of oak or elm but,
to an increasing extent, softwood timber came to be used.

   In medieval buildings floorboards were often laid
lengthways with the joists, housed into the upper surface
so that both boards and joists were visible. Occasionally
floorboards were laid crossways with the joists but with
ends housed in a rebate in the joists. Normally, however,
floorboards are laid across the joists, concealing them
completely. (d90, 91)

   Floorboards may be butt-edged, be joined by loose
tongues in grooves or have integral tongues and grooves.
Usually the ends of floorboards are butted against each
other but in very high quality work they may be connected by loose double-ended dovetailed slips of timber or be
shaped into pointed comb ends which mesh into each
other (forked joint). Nailing was through the boards to
the joists with nail heads punched in, or, as secret nailing,
through tongues. (d92, 93, 94, 95, 96)

   In good quality work superior floorboarding was
applied to a sub-floor of ordinary floorboards nailed to the
joists. (d97)

FLOORING (NAKED FLOORING): the timbers which support
the floorboards (and sometimes the ceiling) of an intermediate floor constitute the flooring, formerly called the
naked flooring or the carcass flooring.

   Floors may be: single floors, that is of joists only; double

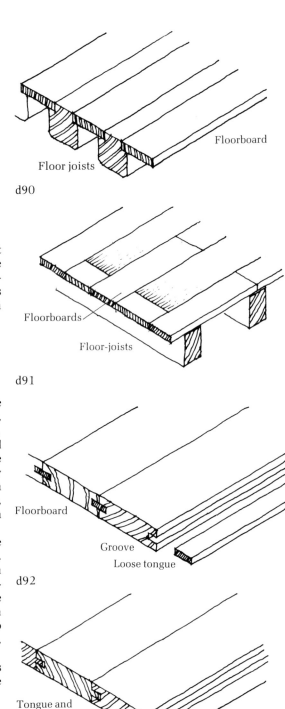

d90

d91

d92

d93

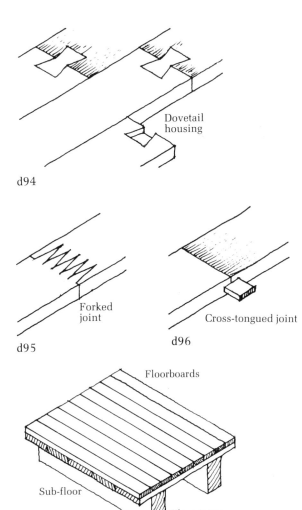

d94

Dovetail housing

Forked joint

d95

Cross-tongued joint

d96

Floorboards

Sub-floor

Floor-joists

d97

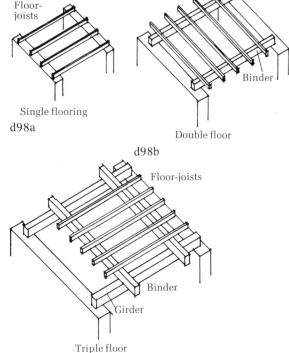

Single flooring

d98a

Floor-joists

Floor-joists

Binder

Double floor

d98b

Floor-joists

Binder

Girder

Triple floor

d98c

floors—with joists carried on binders; or framed floors of girders receiving binders which carry joists (d98a,b,c)

FLYING BRESSUMMER : *see* BRESSUMMER

FOLDING WEDGES : *see* WEDGE

FORELOCK BOLTS : *see* FASTENINGS

FOXTAIL WEDGING : *see* WEDGE

FRAME, FRAMING : the assembly of timbers with the aid of joints and triangulating members to make a self-sustaining entity.

GABLE, GABLED ROOF : *see* ROOF SHAPES

GANG NAIL ROOF TRUSS : a type of trussed rafter roof now widely used and making use of gang nail plates to make joint connections.

Precisely sawn lightweight timber members are placed in a jig, butt-jointed together and then secured by the application to both sides of the joint of a pattern of nail-like projections stamped from the gang nail plate. (d99)

The trussed rafters are placed at normal common rafter spacings along the length of the roof and secured by means of some diagonal or passing brace against collapse. Gang nail roof trusses may also be designed as asymmetrical trusses or to allow for use of the roof space. (d100)

GIBS AND COTTERS : *see* FASTENINGS

GIRDER (GIRDING BEAM)
1. a heavy timber beam carrying binders in a framed triple floor. (d98c)
2. a heavy timber beam capable of carrying a considerable weight between two points of support—carrying, for instance, an internal partition on a floor above.

A girder may be *single*, that is a single piece of timber, or *framed*, that is made up of two or more pieces of timber which have been assembled horizontally (making use of scarfing techniques), or assembled vertically as a flitched beam. A framed girder may be a trussed or composite beam (q.v.).

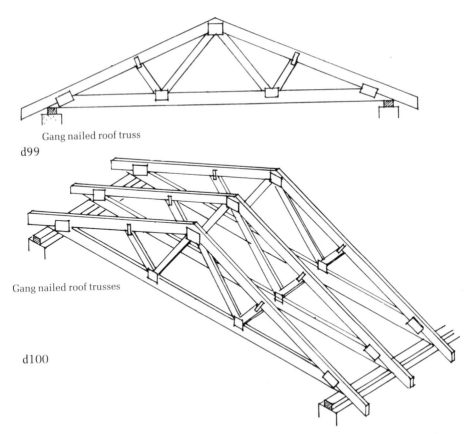

Gang nailed roof truss

d99

Gang nailed roof trusses

d100

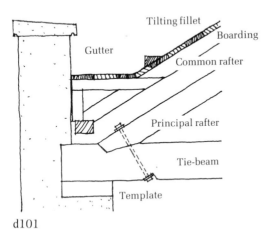

Tilting fillet

Boarding

Gutter

Common rafter

Principal rafter

Tie-beam

Template

d101

GIRTH (GIRT, END GIRT, SIDE GIRT): a timber beam placed horizontally in a wall frame at a level between the sill and the tie-beam or wall-plate. An END GIRT is found in an end or gable wall, a SIDE GIRT in a front or back wall while a CHIMNEY GIRT trims round a fireplace opening. Normally the girts are framed into wall-posts while they may receive

studs as part of the wall construction and joists as part of the floor construction. The term was commonly used in North America from the seventeenth century to the nineteenth century but has also been used in England.

GRILLAGE: a framework consisting of two or more layers of heavy timbers running in opposite directions and serving to carry walls or piers. A grillage is designed to distribute exceptionally heavy loads in ordinary ground or to carry ordinary loads on poor ground.

GROUND PLATE, GROUND SILL: *see* SILL

GROUNDS: battens built into solid walls to act as fixings for panelling or other decorative linings.

GUTTER: a channel to remove rainwater. A carpenter may form a concealed gutter of boarding on battens behind a parapet. (d101)

HALF-HIPPED ROOF: *see* ROOF SHAPES

HALF-TIMBER: various meanings have been attached to this term.
1. A timber frame in which the framing members and panels are exposed.

2. A timber frame based on halved timbers.

3. The use of a timber frame in the upper half of a building with a solid wall in the lower half.

4. Timber framing making use of closely-spaced studs in which approximately equal proportions of timber and plaster are exposed.

HALF TRUSS: part of a truss, usually arranged as half of a full truss, employed at the hipped end of a building. A half king post truss is an example. End crucks are a type of half truss. (d102)

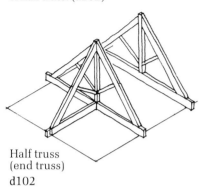

Half truss
(end truss)
d102

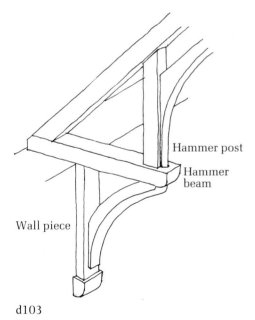

Hammer post

Hammer beam

Wall piece

d103

HAMMER BEAM: a short beam receiving the end of a principal rafter and projecting into a roof space to receive a hammer post and, usually, an arch brace. The hammer beam is supported in turn by an inclined or curved brace. (d103) (**64**)

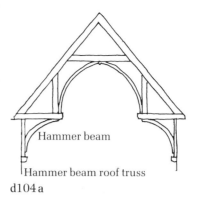

Hammer beam

Hammer beam roof truss
d104a

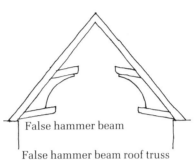

False hammer beam

False hammer beam roof truss
d104b

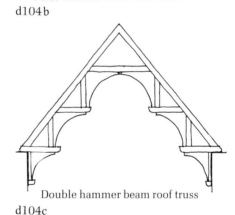

Double hammer beam roof truss
d104c

FALSE HAMMER BEAM: a short bracket extending inwards from a principal rafter but carrying no hammer post. (**64**)

HAMMER-BEAM ROOF: a roof based on the use of hammer-beam roof trusses.

HAMMER-BEAM ROOF TRUSS: a roof truss making use of hammer beams and hammer posts with their attendant braces and arch braces. The hammer-beam roof truss has the appearance of a tie-beam truss with the central part of the tie-beam cut away; the hammer beams are in effect the remaining ends of the tie-beam. (d104a)

FALSE HAMMER-BEAM ROOF TRUSS: a roof truss having false hammer beams at one or more levels. (d104b)

DOUBLE HAMMER-BEAM ROOF TRUSS: a hammer-beam roof truss with an extra pair of hammer beams at a high level carrying hammer posts. (d104c)

HAMMER POST: a stout post rising from the inner end of a hammer beam to meet the principal rafter. Usually the bottom of the post is tenoned into the upper surface of the hammer beam, but alternatively the end of the hammer beam is tenoned into the face of the hammer post whose bottom end is then made into a pendant. (d103)

HANGER: a light suspending member dropping from rafter or purlin to carry ceiling joists by means of a horizontal runner. (d105)

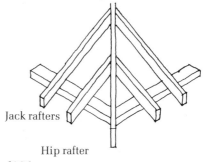

Jack rafters

Hip rafter

d106

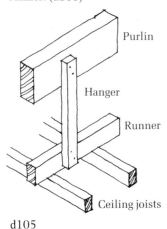

Purlin

Hanger

Runner

Ceiling joists

d105

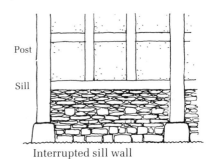

Post

Sill

Interrupted sill wall

d107

the studs rise from a timber sill which rests on a plinth wall but runs *between* posts into which the sill is tenoned. Usually posts and studs rise from the same sill. (d107) (63)

HANGING KNEE: *see* KNEE

HARDWOOD: *see* TIMBER

HERRINGBONE STRUTTING: *see* STRUTTING

HEWN KNEE: *see* KNEE

HIPPED ROOF: *see* ROOF SHAPES

HIP RAFTER: a deep timber member following the external angle at the junction of two slopes of a roof. The hip rafter runs from the junction of wall-plates to the ridge (or to meet other hip rafters in a pyramidal roof) and receives the ends of shortened or jack rafters. (d106)

HOOK PINS: *see* PINS *under* PEGS

HORIZONTAL RAFTERS: *see* COMMON PURLINS *under* PURLINS

INTERRUPTED SILL WALL: a timber-framed wall in which

63. Interrupted sill, house at Thornhill Lees, Yorks. W.R.

64. Hammer beam, Church of St. Wendreda, March, Cambs.

134

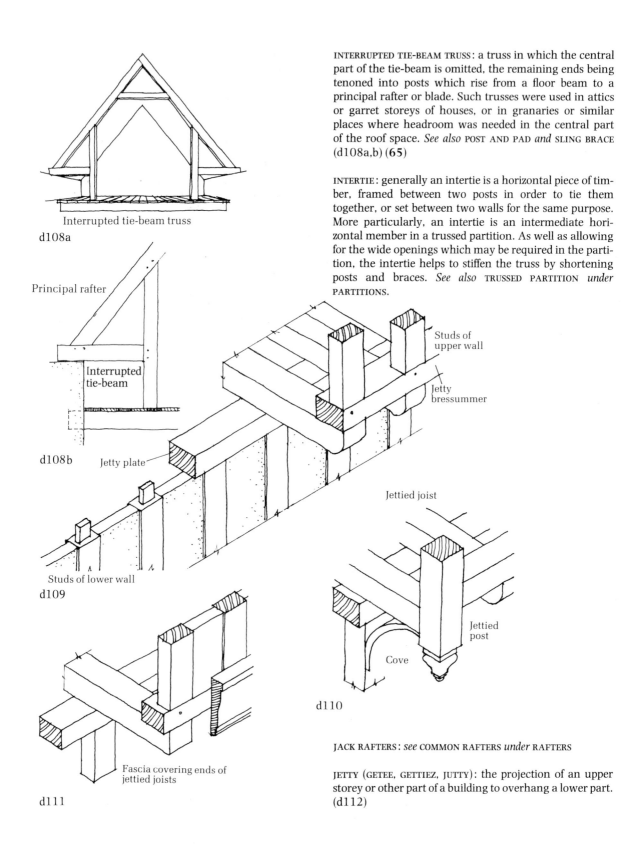

Interrupted tie-beam truss

d108a

INTERRUPTED TIE-BEAM TRUSS: a truss in which the central part of the tie-beam is omitted, the remaining ends being tenoned into posts which rise from a floor beam to a principal rafter or blade. Such trusses were used in attics or garret storeys of houses, or in granaries or similar places where headroom was needed in the central part of the roof space. *See also* POST AND PAD *and* SLING BRACE (d108a,b) (**65**)

INTERTIE: generally an intertie is a horizontal piece of timber, framed between two posts in order to tie them together, or set between two walls for the same purpose. More particularly, an intertie is an intermediate horizontal member in a trussed partition. As well as allowing for the wide openings which may be required in the partition, the intertie helps to stiffen the truss by shortening posts and braces. *See also* TRUSSED PARTITION *under* PARTITIONS.

Principal rafter

Interrupted tie-beam

d108b   Jetty plate

Studs of upper wall

Jetty bressummer

Jettied joist

Studs of lower wall

d109

Jettied post

Cove

d110

Fascia covering ends of jettied joists

d111

JACK RAFTERS: *see* COMMON RAFTERS *under* RAFTERS

JETTY (GETEE, GETTIEZ, JUTTY): the projection of an upper storey or other part of a building to overhang a lower part. (d112)

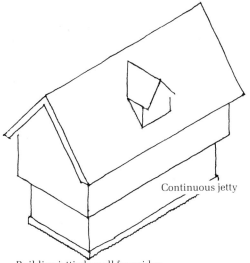

Continuous jetty

Building jettied on all four sides

d112

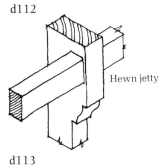

Hewn jetty

d113

65. Interrupted tie-beam, Banister Hall, Walton-le-Dale, Lancs.

In the most common type the studs of a wall are framed into a plate or JETTY PLATE. Floor-joists are then jettied or projected and on the ends of these a sill or JETTY BRESSUMMER is placed from which rise the studs of the upper or jettied wall. In a less common form beams or side girts are projected and upper wall-posts are jointed into their ends; a beam running between the upper wall posts carries the feet of the studs of the upper or jettied wall. (d109, d110) (**66**) (**68**)

The gable of a wall or dormer may also be jettied.

The ends of projecting joists may be hidden behind an applied decorative fascia while the junction between the underside of projecting joists and the lower wall may be masked by a cove of timber and plaster. (d111)

CONTINUOUS JETTY: a jettied storey or other part of a building running along the whole length of a wall. (**67**)

FALSE JETTY (HEWN JETTY): the slight projection of one storey from that below achieved by cutting back the lower parts of posts which run through more than one storey. (d113)

JOGGLE PIECE: *see* KING POST

JOINER: a craftsman skilled in preparing and joining together or otherwise assembling wood for panelling, screenwork or other fittings or for furniture.

JOINTED CRUCKS: *see* CRUCKS

JOINTS: the details whereby one timber member is related, attached or linked to another.

BRIDLE JOINT: used where one member runs into another, usually at an oblique angle, and the receiving member has two rebates cut away leaving the centre part intact and standing proud ready to be engaged in a recess cut into the other member. (d114a)

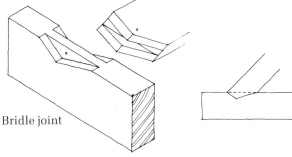

Bridle joint

d114a

COGGED JOINT: used when one member, such as a beam, is resting on another member such as a wall-plate; the upper member has a sinking taken out to correspond

66. Jettying, St. John's Alley, Devizes, Wilts.

68. Jettying, Myddleton Place, Saffron Walden, Essex

67. Jettying, cottages at Cerne Abbas, Dorset

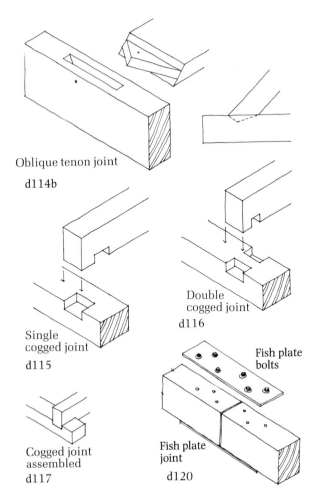

Oblique tenon joint
d114b

Single
cogged joint
d115

Double
cogged joint
d116

Cogged joint
assembled
d117

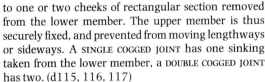

Fish plate
bolts

Fish plate
joint
d120

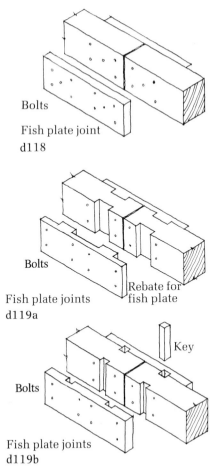

Bolts

Fish plate joint
d118

Bolts

Rebate for
fish plate

Fish plate joints
d119a

Bolts

Key

Fish plate joints
d119b

to one or two cheeks of rectangular section removed from the lower member. The upper member is thus securely fixed, and prevented from moving lengthways or sideways. A SINGLE COGGED JOINT has one sinking taken from the lower member, a DOUBLE COGGED JOINT has two. (d115, 116, 117)

FISHED JOINT, FISH PLATE JOINT: used when one member has to be lengthened with the aid of another of similar cross-section butted end to end. Thin timber or metal pieces are placed side by side (usually) or above and below (occasionally) the butt-joint and bolted through. To help prevent the wooden fish plates from sliding along the members to be joined, they are sometimes rebated to relate to corresponding members in the main joint; alternatively both fish plates and main members may be cut to receive hardwood keys. (d118, 119a, b 120)

HALVED JOINT: used when one member crosses another as when a collar crosses a rafter. A cut extending to half the depth of one member fits into a rebate cut to

half the depth of the other member. Both members are correspondingly weakened but uniform surfaces on both sides are maintained. In the DOVETAILED HALVED JOINT, one member has a SINGLE or DOUBLE sided dovetail shape to one half so as to engage in a rebate of similar shape in the other. A LATERAL BEVELLED HALVED JOINT has diminishing cross-section while a LONGITUDINAL BEVELLED HALVED JOINT diminishes along its length. (d121, 122, 123, 124, 125)

HOUSED JOINT: used where one member runs into another member maintaining a common upper surface, as when a joist meets a binder. In the SIMPLE HOUSED JOINT the lighter member has cross-section unaltered and slots into a recess cut in the heavier member, the lighter member is thus prevented from moving sideways but may slip out of the joint lengthways. In the DOVETAIL HOUSED JOINT the end of the lighter member is tapered back in dovetail shape to resist withdrawal. A NOTCHED HOUSED JOINT achieves a similar effect with the use of a triangular projection

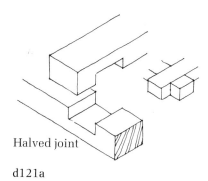

Halved joint

d121a

Halved joint

d121b

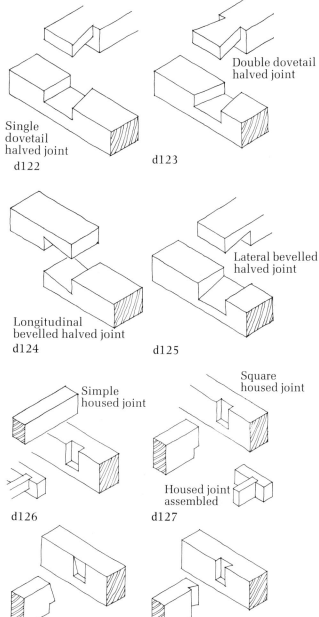

Single
dovetail
halved joint

d122

Double dovetail
halved joint

d123

Longitudinal
bevelled halved joint

d124

Lateral bevelled
halved joint

d125

Simple
housed joint

d126

Square
housed joint

Housed joint
assembled

d127

Bevelled housed joint

d128

Dovetail housed joint

d129

69. Jointing, barn at Grange Farm, Coggeshall, Essex

from the heavier member engaging in a corresponding slot in the face of the lighter member. To avoid cutting away too much of the heavier member the housing may only extend part way down in a SQUARE HOUSED JOINT or this partial housing may be tapered back to preserve the top of the timber intact as BEVELLED HOUSING. (d126, 127, 128, 129, 130)

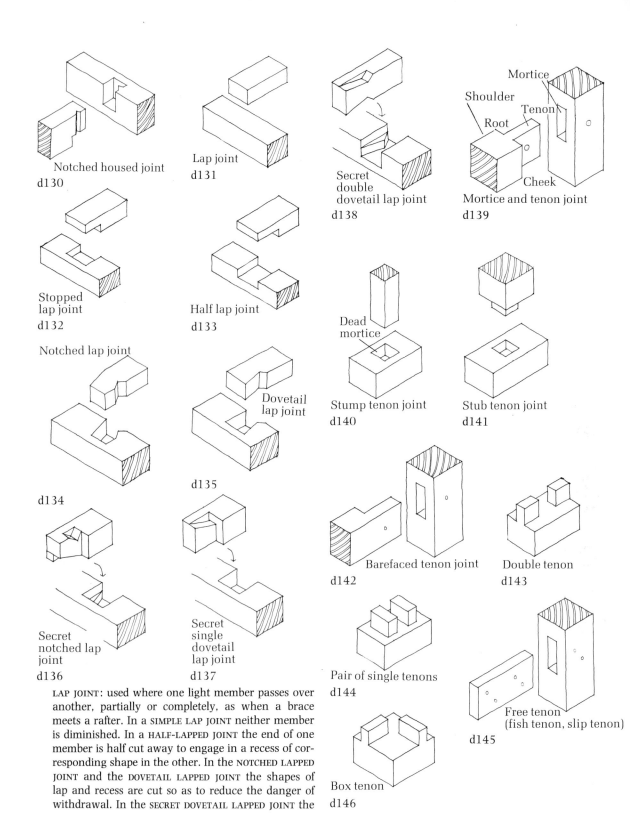

Notched housed joint
d130

Lap joint
d131

Secret double dovetail lap joint
d138

Mortice and tenon joint
d139

Shoulder
Mortice
Root
Tenon
Cheek

Stopped lap joint
d132

Half lap joint
d133

Dead mortice

Stump tenon joint
d140

Stub tenon joint
d141

Notched lap joint
d134

Dovetail lap joint
d135

Secret notched lap joint
d136

Secret single dovetail lap joint
d137

Barefaced tenon joint
d142

Double tenon
d143

Pair of single tenons
d144

Box tenon
d146

Free tenon (fish tenon, slip tenon)
d145

LAP JOINT: used where one light member passes over another, partially or completely, as when a brace meets a rafter. In a SIMPLE LAP JOINT neither member is diminished. In a HALF-LAPPED JOINT the end of one member is half cut away to engage in a recess of corresponding shape in the other. In the NOTCHED LAPPED JOINT and the DOVETAIL LAPPED JOINT the shapes of lap and recess are cut so as to reduce the danger of withdrawal. In the SECRET DOVETAIL LAPPED JOINT the

142

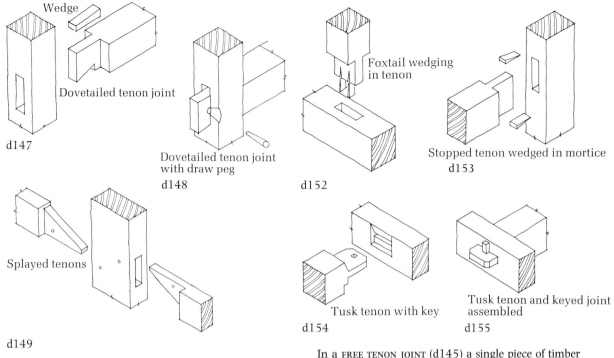

Wedge

Dovetailed tenon joint

d147

Dovetailed tenon joint
with draw peg
d148

Splayed tenons

d149

Face
halved
tenons

Mortice for
chase tenon

d150

d151

Foxtail wedging
in tenon

d152

Stopped tenon wedged in mortice
d153

Tusk tenon with key

d154

Tusk tenon and keyed joint
assembled
d155

shaping is confined to the bottom half of the lap so that on the surface the joint appears to be a half-lapped joint. (d131, 132, 133, 134, 135, 136, 137, 138)(**69**)

MORTICE AND TENON JOINT: the end of one member, the tenon, is cut away to form one or two shoulders and engages, normally at right angles, with a sinking, the mortice, of corresponding depth and cut part way or completely through another member. (d139) In the BAREFACED TENON JOINT (d142) only one part is cut away giving only one shoulder. A DOVETAIL TENON JOINT is used when a member is to be joined to one which is already in position, the mortice is cut as a chase and the tenon is given a dovetail shape so that it cannot withdraw when put in position. (d147, 148)

In a FREE TENON JOINT (d145) a single piece of timber acts as a double-ended tenon and engages in two mortices. A STUB-TENON (d141) has a short fat tenon cut in the end of a post. It acts as a locating joint as does the STUMP TENON JOINT (d140) in which the bottom end of the post is not cut away at all but engages in a dead mortice passing a short distance into a sill. In a TUSK TENON joint (d154, 155) used mostly in floor construction, the tenon passes through the mortice hole and projects so that it can be tightly wedged in position. The upper cheek of the tenon is tapered back so as to reduce the timber cut away from beside the mortice. FACE HALVED and SPLAYED TENONS (d149, 150) were used when two members were passing through the same mortice hole, as when rails engaged with a slender post; the tenons were shaped to pass each other within the single depth of the mortice. The OBLIQUE TENON is occasionally used when one member joins another at an angle, as when a principal rafter rises from the end of a tie-beam. The oblique tenon is an alternative for a bridle joint in such a situation. (d114b)

NOTCHED JOINT: used in the same circumstances as a cogged joint which it resembles except that a recess is cut in the underside of the upper member of the same width as the upper surface of the lower member. In a DOUBLE NOTCHED JOINT there are recesses cut into both upper and lower members. (d156, 157, 158)

SCARFED JOINTS: used when a long member is to be

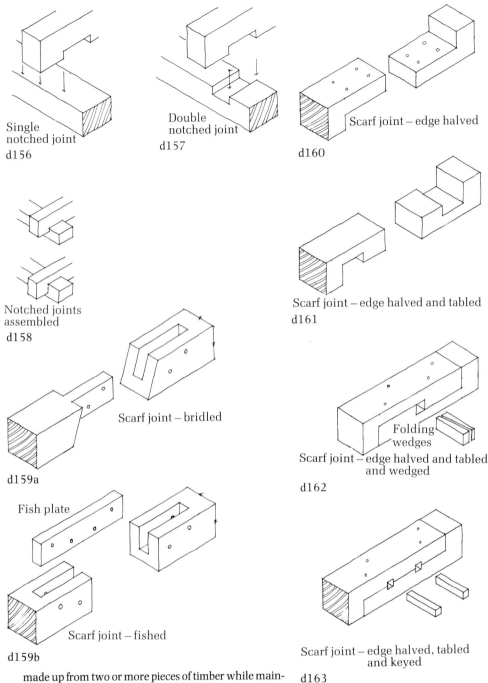

Single
notched joint
d156

Double
notched joint
d157

Scarf joint – edge halved
d160

Notched joints
assembled
d158

Scarf joint – edge halved and tabled
d161

Scarf joint – bridled

d159a

Scarf joint – edge halved and tabled
and wedged
d162

Folding
wedges

Fish plate

Scarf joint – fished
d159b

Scarf joint – edge halved, tabled
and keyed
d163

made up from two or more pieces of timber while main-taining a uniform cross-section. The variations of scar-fed joints are BRIDLED, EDGE HALVED, FACE HALVED, FISHED and SPLAYED. All are pegged together as well as fitting together in various ways.

The BRIDLED SCARF JOINT is a special sort of bridled joint (q.v.) in which a tenon projecting from one member

engages in an open-topped mortice in the other; the abutments may be vertical or splayed. (d159a)

EDGE HALVED SCARF JOINT. This has the top of one member cut away on the square to overlap the bottom of

144

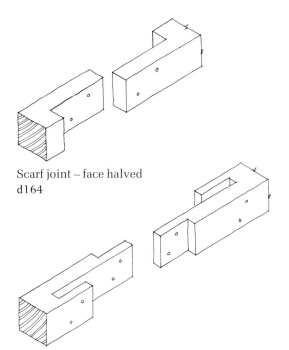

Scarf joint – face halved
d164

Scarf joint – face halved with bladed abutments
d165

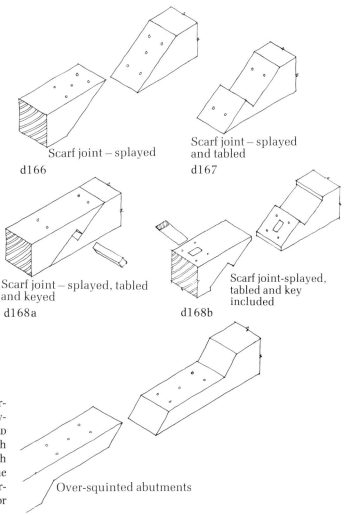

Scarf joint – splayed
d166

Scarf joint – splayed and tabled
d167

Scarf joint – splayed, tabled and keyed
d168a

Scarf joint-splayed, tabled and key included
d168b

Over-squinted abutments

d169

the other member which has been cut away to correspond. (d160) The meeting surface of the edge halving may be further cut away to form an EDGE HALVED AND TABLED JOINT either with simple tabling or with tabling tightened by a pair of folding wedges, or with the tabling cut to allow the insertion of keys. The abutments may be vertical or splayed outwards (over-squinted) or splayed inwards (under-squinted) or bridled as in a bridled scarf joint. (d161, 162, 163)

FACE HALVED SCARF JOINT. This has the side of one member cut away on the square to lap with the side of the other member which has been cut away to correspond. Abutments are normally vertical but in the bladed abutment rebates like miniature face halvings are formed. (d164,165). The FISHED SCARF JOINT is as a bridled scarf joint but with a loose fish plate engaging in two open-ended mortices. (d159b)

The SPLAYED SCARF JOINT (d166) is one in which the two members are cut on the horizontal splay so that the top of one member overlaps the bottom of another. In the SPLAYED AND TABLED SCARF JOINT (d167) the splay is in two interlocking parts and provision may be made for a key which tightens the joint. (d168) Apart from a simple abutment in which the splay dies away, the abutment may have a short vertical surface or a short surface inclined inwards as a square under-squinted

abutment. (70) Alternatively either of these abutments may be bridled as in a bridled scarf joint or may be sallied i.e. have a projection shaped like the prow of a ship meeting a corresponding sinking. In a more complicated variation the splayed joint, with its vertical or under-squinted abutments, is further connected by a tenon projecting from half of one splay engaging in a mortice cut from half the other splay. (d169, 170, 171, 172, 173, 174, 176, 177, 178)

SCISSORS JOINT: a type of scarfed joint used vertically as when a post is being formed out of two members or (more commonly) a rotten part of a post is being replaced. The joint is like a double splayed scarf joint, the splay of one half running in the opposite direction to the splay in the other half. (d175)

70. Scarf joint, North Warehouse, Gloucester

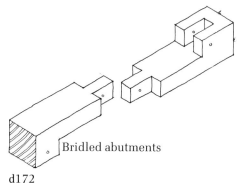

Bridled abutments

d172

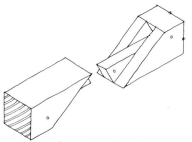

Scarf joint – splayed, counter-tongued and grooved

d173

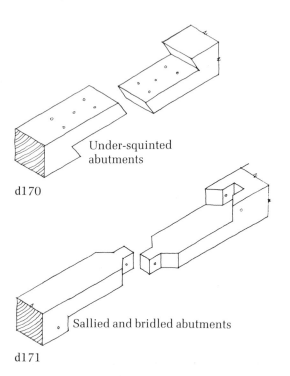

Under-squinted abutments

d170

Sallied and bridled abutments

d171

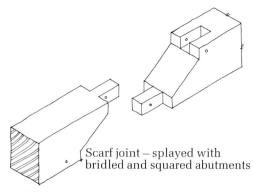

Scarf joint – splayed with bridled and squared abutments

d174

TRENCHED JOINT: a variation of the notched joint in which one member runs across and is carried by another member, a recess being cut out of the lower member to receive the bottom part of the upper member. (d179)

JOISTS (JOYCE, GOICE): the horizontal timbers which support floorboards or carry boarded or plastered ceilings. Joists are of small scantling and ceiling joists are usually smaller than floor-joists. Although there are some exceptions, early joists were approximately square in section while later joists were deeper than they were wide. Joists have usually been laid in parallel and between about 12 ins (305 mm) and 24 ins (610 mm) apart. (d180, 181, 183)

Openings in joisted floors as for staircases or around chimney stacks are made by TRIMMING the joists i.e. cutting off some and carrying them on short joists acting as beams.

BINDING JOIST: an alternative term for a SIDE GIRT.

146

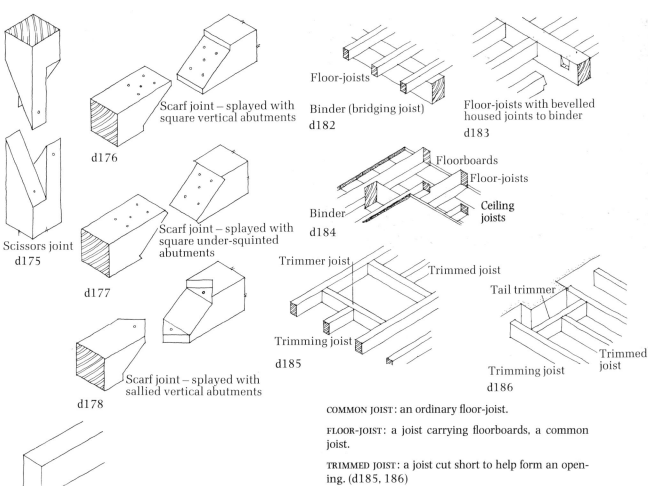

Scissors joint
d175

Scarf joint – splayed with
square vertical abutments
d176

Scarf joint – splayed with
square under-squinted
abutments
d177

Scarf joint – splayed with
sallied vertical abutments
d178

Trenched joint
d179

Floor-joist – upright
d180

Floor-joist
square section
d181

Floor-joists
Binder (bridging joist)
d182

Floor-joists with bevelled
housed joints to binder
d183

Floorboards
Floor-joists
Binder
Ceiling
joists
d184

Trimmer joist
Trimmed joist
Trimming joist
d185

Trimmed joist
Tail trimmer
Trimming joist
Trimmed joist
d186

COMMON JOIST: an ordinary floor-joist.

FLOOR-JOIST: a joist carrying floorboards, a common joist.

TRIMMED JOIST: a joist cut short to help form an opening. (d185, 186)

TRIMMER JOIST: a joist (which may be stouter than other joists) which spans between trimming joists and helps to make an opening in a floor. A TAIL TRIMMER is specifically a trimmer joist, near a wall, and receiving the ends of joists to avoid any flues. (d185, 186)

TRIMMING JOIST: a joist (which also may be stouter than other joists) which receives the ends of trimmer joists as an opening is formed in the floor. (d185, 186)

JOWL (GUNSTOCK): the enlarged head of a post. A jowl permits a member carried by the post to be properly located and securely fixed. A jowl is usually found at the head of a structural post, such as a wall-post or arcade post, but may be found at the head of a crown post. The shape of the jowl varies from a graceful flare to a distinct and square-cut projection. GUNSTOCK is an alternative term used especially in North America for a jowl. (d187, 188a, b, c, d)

JUFFERS: timbers about 4 ins (100 mm) or 5 ins (127 mm)

BRIDGING JOIST: a beam providing intermediate support to floor-joists. See BINDER. There is some confusion in that in certain text books the term 'Bridging Joist' is used for an ordinary floor-joist. (d182)

CEILING JOIST: a joist separate from a floor-joist and carrying a boarded or plastered ceiling. (d184)

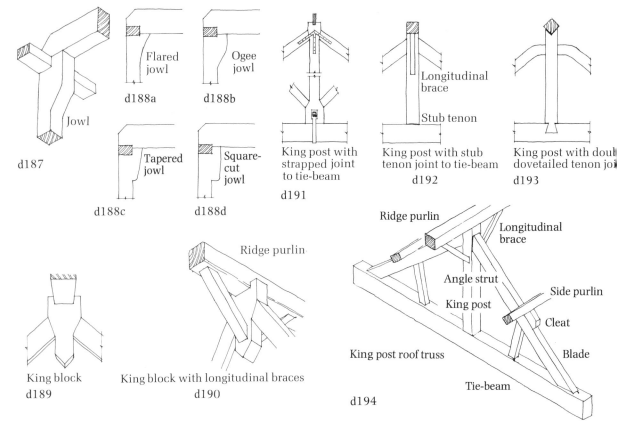

Jowl

d187

Flared jowl
d188a

Ogee jowl
d188b

Tapered jowl
d188c

Square-cut jowl
d188d

King post with strapped joint to tie-beam
d191

King post with stub tenon joint to tie-beam
d192

King post with doub[le] dovetailed tenon joi[nt]
d193

Ridge purlin

King block
d189

King block with longitudinal braces
d190

Ridge purlin

Longitudinal brace

Angle strut

Side purlin

King post

Cleat

Blade

King post roof truss

Tie-beam

d194

square and of several lengths; a term noted as obsolete in the early nineteenth century.

KERB PRINCIPALS: *see* PRINCIPAL RAFTERS

KING BLOCK: a block of timber set between blades at the apex of a roof truss. Usually a ridge purlin is carried by the king block and the block may be deep enough to carry the end of a brace rising to the ridge purlin. (d189,190)

KING BOLT: a wrought-iron rod used in place of a king post. (d196)

KING PENDANT: a term sometimes used for a deep king block.

KING POST (KING PIECE, JOGGLE PIECE): a vertical timber rising from the centre of a tie-beam to carry a ridge purlin or ridge. There may be longitudinal braces rising from the king post to the ridge purlin. In the North of England, and

71. King post roof truss, Dorchester, Dorset.

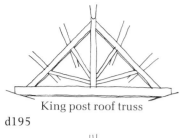

King post roof truss

d195

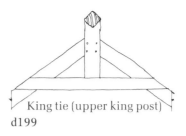

King tie (upper king post)

d199

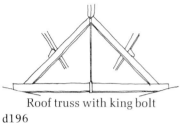

Roof truss with king bolt

d196

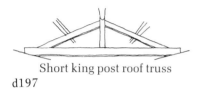

Short king post roof truss

d197

No ridge member

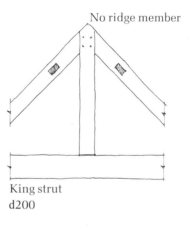

King strut

d200

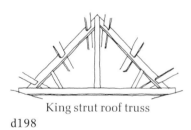

King strut roof truss

d198

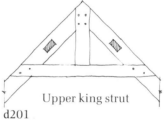

Upper king strut

d201

especially in exposed gables the king post may either receive braces (rather like inclined studs) rising from each side of the tie-beam or it may provide a seating for such members rising to the blades of the truss. Generally, and especially in later examples, a single pair of braces rose from near the base of the king post to meet the blades.

Early king posts were located on the top of the tie-beam with the aid of a stub-tenon, later king posts recognised as tension or suspending members were linked to the tie-beam by a dovetail lapped joint, by a bolt or by a strap. (d191, 192, 193)

Some glossaries of the eighteenth and nineteenth centuries give crown post as an alternative term for king post, but nowadays the term crown post is reserved for those central posts in a roof which rise from a tie-beam but do not reach the ridge.

KING POST ROOF TRUSS: a roof truss consisting essentially of a tie-beam, a pair of principal rafters or blades and a king post. The king post rises from the tie-beam to support a ridge, the principal rafters or blades are inclined and run from near the ends of the tie-beam to be received by the king post near its head and they carry side purlins. Subsidiary members may run from king post or tie-beam to the principal rafters or blades. (d194, 195) (71) (72)

KING STRUT: a member rising like a king post vertically from a tie-beam towards the apex of a roof receiving the ends of principal rafters or blades but not carrying a ridge. An UPPER KING STRUT is like a king strut but rises from a collar rather than a tie-beam. (d200, 201)

KING STRUT ROOF TRUSS: a roof truss consisting of a tie-beam and two inclined principal rafters or blades meeting at the apex and receiving a king strut. There may also be subsidiary members rising from the king strut or the tie-beam to the blades. (d198, 200)

72. King post roof truss, Howard Street Warehouse, Shrewsbury, Salop

KING TIE (UPPER KING POST): a short member which runs vertically from the centre of a collar or straining beam, receives the upper ends of principal rafters or blades and carries a ridge or ridge purlin. It may also send braces to the ridge purlin. (d199)

SHORT KING POST: a king post used in a low-pitched roof and receiving the ends of only slightly sloping principal rafters or blades. (d197)

KNEE: a sharply angled or curved piece of timber resembling a leg bent at the knee. Such timbers are usually attached to other structural members as reinforcement. Thus a HANGING KNEE is set below a joint, a LODGING KNEE runs horizontally at a joint and a STANDING KNEE is set above a joint. The terms may derive from shipwrights' work in which knees were frequently used. (d202)

HEWN KNEE: a timber cut or hewn so as to give a knee shape at one end. (d203)

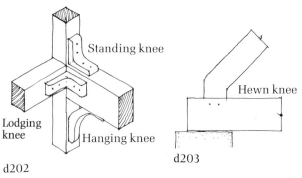

KNEE BRACE: *see* BRACE

KNEE PIECE: *see* BASE CRUCK

LAMINATED TIMBER: layers of wood fastened together by bolting, screwing, or nailing, but most commonly by glueing. The characteristics of laminated timber are known or can be predicted more accurately than those of natural timber. Given cross sections of laminated timber are usually stronger than those of natural timber. Lamination is also used as a way of curving timber to a desired shape

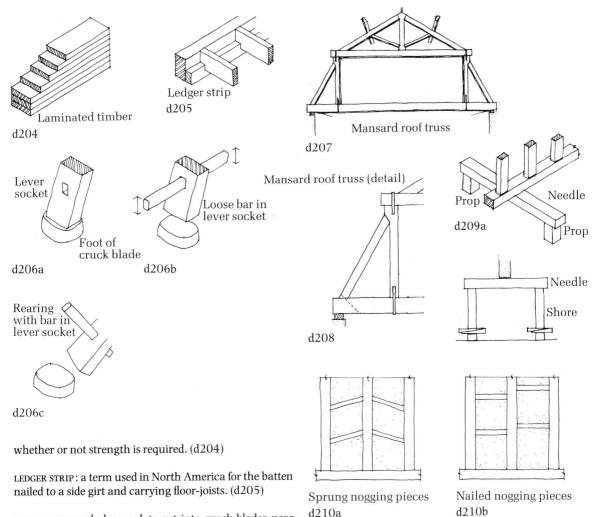

Laminated timber
d204

Ledger strip
d205

Mansard roof truss
d207

Lever socket
d206a

Loose bar in lever socket

Foot of cruck blade
d206b

Mansard roof truss (detail)

Prop

Needle

d209a

Prop

Rearing with bar in lever socket

d206c

d208

Needle

Shore

Sprung nogging pieces
d210a

Nailed nogging pieces
d210b

whether or not strength is required. (d204)

LEDGER STRIP: a term used in North America for the batten nailed to a side girt and carrying floor-joists. (d205)

LEVER SOCKETS: holes or slots cut into cruck blades near their feet. They were intended to receive a lever which helped to make the first lift possible in rearing a cruck truss, and also to receive a beam or iron bar with which the foot of a cruck could be manoeuvred into its final position. (d206a, b, c)

LOGS: trunks of felled trees with the head topped and the branches lopped off.

LUMBER: the North American term for logs and ultimately for structural timber.

MANSARD ROOF *and* HIPPED MANSARD ROOF: *see* ROOF SHAPES

MANSARD ROOF TRUSS: the truss designed to allow usable roof space within a mansard roof of two pitches. There are various versions of such a truss but one type commonly used resembles a king post roof truss raised above

a queen post truss, the straining beam of the lower truss acting as the tie-beam of the upper. (d207, 208)

MATHEMATICAL TILE CLADDING: *see* CLADDING

MIDRAIL (MIDDLE RAIL): *see* PANELS

MORTICE: *see* MORTICE AND TENON JOINT *under* JOINTS

M-SHAPED ROOF: *see* ROOF SHAPES

MUNTIN AND PLANK PARTITION: *see* PARTITIONS

NEEDLE: a short timber, wrought-iron or steel beam passed through a hole in a wall to give temporary support as in underpinning. In small-scale work the needle is supported on short props while in larger scale work needles are

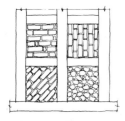

Brick, stone, flint nogging
d211

carried on dead shores, that is stout posts braced and secured with the aid of iron dogs. (d209a, b)

NOGS (NOGGINGS, NOGGING PIECES):
1. short lengths of timber of about 4 ins (100 mm) by 3 ins (76 mm) scantling set at vertical intervals of about 3 ft (914 mm) or 4 ft (1219 mm) between studs to stiffen a stud partition. Usually the noggings are sprung or wedged into position though they may be nailed in more recent work. Noggings differ from rails, which are jointed to studs or posts. (d210a, b)
2. the solid infill used in panels in half-timber work as an alternative to or replacement of wattle and daub. Brick nogging is the most common version though stone and flint nogging have also been used. (d211) (74)

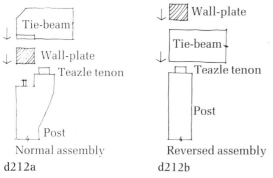

Normal assembly
d212a

Reversed assembly
d212b

NORMAL ASSEMBLY: the term coined to apply to the detail most commonly found at the head of a wall-post or aisle post whereby the wall-plate or arcade plate rests on the post and the tie-beam then rests on the plate or the plate and post combined. This detail is different from that often found, for instance, on an aisle wall where a beam rests on the post and then a wall-plate rests on the beam. The term REVERSED ASSEMBLY has been coined for this. (d212a, b)

OPEN TRUSS: a truss lacking a tie-beam. The arch-braced collar beam truss is probably the most common type. Open trusses are used where the tie-beam would be unsightly or inconvenient. See also OPEN CRUCK TRUSS under CRUCK TRUSS.

PAD: a short length of timber receiving the end of a principal rafter or blade and running into the roof space (that is across rather than along a wall). See also POST AND PAD. (d246)

PAN: a North Country term for a wall-plate. See also POST AND PAN

PANELS: the areas between the studs or posts and the rails or other horizontal members in timber-framed walls. Most commonly panels were filled with wattle and daub but might be filled with slabs of brick, stone, slate, tile or with nogging. Panels may be tall and narrow or approximately square or they may be of some shape intended to be ornamental. (d213a, b, c, 214, 215, 216, 217, 218) (73)

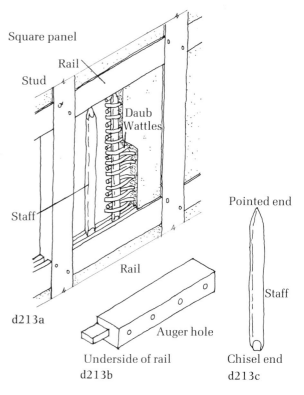

d213a

Underside of rail
d213b

Chisel end
d213c

CLOSE STUDDING: narrow panels between closely-spaced studs and normally a full storey in height. Often stud and panel are of roughly similar width. (d219)

CLOSE STUDDING WITH MIDDLE RAIL: as close studding but with short lengths of rail running between studs at about mid-height. (d220)

CLOSE PANELLING: the division of a storey of a wall into small squares, three or four in the storey height. (d221)

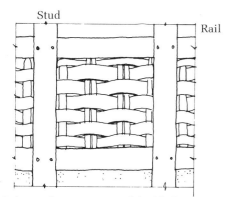

Stud

Rail

Split wattles woven round doubled staves
d214

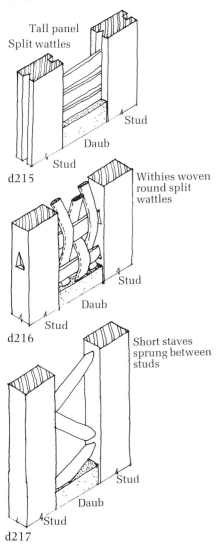

Tall panel
Split wattles

Stud

Daub

Stud

d215

Withies woven
round split
wattles

Stud

Daub

Stud

d216

Short staves
sprung between
studs

Stud

Daub

Stud

d217

73. Boarded infill, Coppice Farm, Ashley, Cheshire

Stone or tile

d218

153

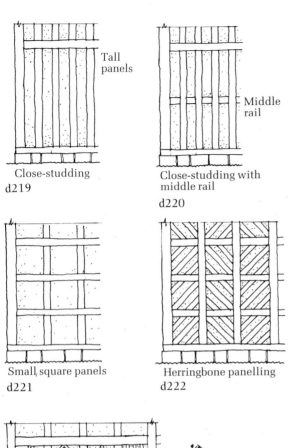

Tall panels

Middle rail

Close-studding
d219

Close-studding with middle rail
d220

Small square panels
d221

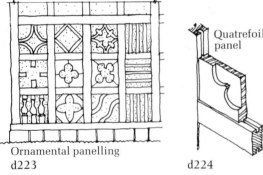

Herringbone panelling
d222

Ornamental panelling
d223

Quatrefoil panel

d224

Large panels, daub on staves
d225

DIAMOND PANELLING: panelling formed into a diamond or lozenge shape; found most often in the gables of timber-framed buildings. (d226)

HERRINGBONE PANELLING: the division of adjacent square panels by means of non-structural diagonal members inclined in opposite directions in each row or column of panels, (d222)

LARGE PANELS: areas of storey height and bay width lacking subdivision by studs or rails. (d225)

ORNAMENTAL PANELLING: panels square or approximately square in shape and divided ornamentally in one of several ways e.g. with short non-structural diagonals cutting off the corners, short concave or convex members at each corner, short pieces running towards the centre of the panel, infill with wooden slabs cut or recessed to give a quatrefoil effect. (d223, d224) (76)

SQUARE PANELS: panels square or approximately so. (75)

PARTITIONS: internal walls dividing a building into usable spaces. Timber-framed partitions are usually found in timber-framed buildings but may also be found in solid-walled buildings.

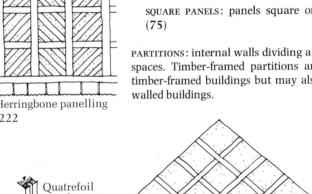

Diamond panelling
d226

Stud

Rail

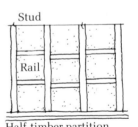

Half-timber partition
d227

HALF-TIMBER PARTITION: a timber-framed partition with framing panels and timber members exposed. (d227, 228a, b)

MUNTIN AND PLANK PARTITION: a partition made of grooved studs or muntins with timber boards or planks let into the grooves. The muntins often have edges chamfered or otherwise decorated. The planks are sometimes tapered in section to fit into the grooves.

74. Bricknogging, Old Manor House, Romsey, Hants.

Usually the muntins are tenoned into a timber sill at the foot and a plate at the head. A muntin and plank partition may be structural, carrying floor-joists, or free-standing and non-structural, supporting only itself. (d229a, b, c) (**77**)

QUARTERED PARTITION (STOOTHED PARTITION, STUD PARTITION): a timber-framed partition incorporating quarters or studs but covered with lath and plaster. 'Stoothins' was a provincial term for studs.

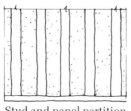

Stud and panel partition
d228a

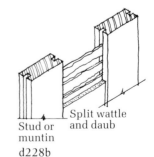

Stud or muntin

Split wattle and daub

d228b

75. Square panelling, Pembridge, Herefords.

TRUSSED PARTITION: a timber-framed load-bearing partition. The members are triangulated so that the partition is self-supporting, that is it carries its own weight across an opening and may also be designed to carry floor or roof loads over openings. A trussed partition is in effect a very deep trussed beam. A trussed partition will incorporate a head, sill, posts, braces and studs and, if with an opening or more than about 10 ft (3m) high, an intertie. The surfaces of the trussed partition are usually covered with lath and plaster concealing the structural members. (d230,231)

PASSING BRACE: *see* BRACE

PEGS (PINS, TRENAILS): short lengths of wood of square or round cross-section, or part one and part the other, driven into bored holes to draw together and secure joints in timber-framing. *See also* DRAWBORING (d232a,b) (78)

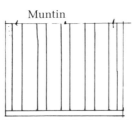

Muntin

Muntin and plank partition
d229a

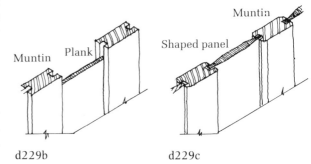

Muntin

Plank

d229b

Muntin

Shaped panel

d229c

156

76. Quatrefoil panelling, Lydiate Hall, Lydiate, Lancs.

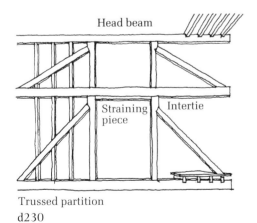

Head beam

Straining piece

Intertie

Trussed partition
d230

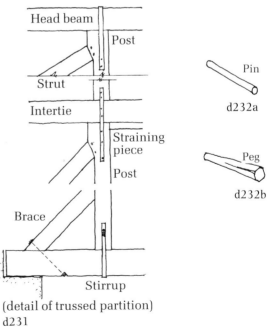

Head beam

Post

Strut

Intertie

Straining piece

Post

Brace

Stirrup

(detail of trussed partition)
d231

Pin

d232a

Peg

d232b

77. Muntin and plank panelling, Berllan-deg, Llanhenog, Mon.

EDGE PEGS are driven through top or bottom of a joint. (d233)

FACE PEGS are driven through the side of a joint. (d234)

HOOK PINS are used in temporary assembly and withdrawn to be replaced by pegs. (d7)

PINS are of cylindrical section and used with mortice and tenon joints. (d232b)

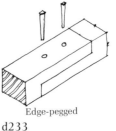

Edge-pegged

d233

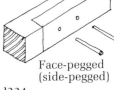

Face-pegged (side-pegged)

d234

PENDANT: a short piece of timber suspended from above and with ornamental treatment at its bottom end. (d235)

PENTISE, PENT ROOF: *see* ROOF SHAPES

PETTRIL (PEYTREL): like pan, an obsolete term for a wall-plate.

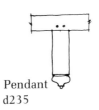

Pendant
d235

Pillow

d236

78. Pegs, barn at Corrimony, Inverness-shire, Scotland

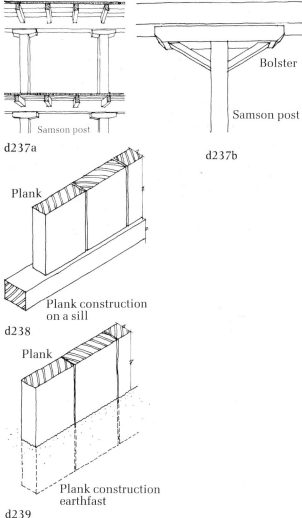

d237a

Bolster

Samson post

d237b

Plank

Plank construction
on a sill

d238

Plank

Plank construction
earthfast

d239

PIGNON, PYNNION, *see* GABLE

PILLOW: a short piece of timber acting as a spreader, collecting loads and concentrating them on to the head of a post. (*See* SAMSON POST.) (d236) A BOLSTER is a longer pillow (d237) (**87**)

PINS: *see* PEGS

PITCH: a black sticky substance, the residue from the distillation of pine or fir. Pitch was heated, liquefied and applied to outside woodwork as a waterproofing material.

PLANK: timber sawn to a cross-section of about 2 ins to 4 ins (51mm to 100 mm) thick and more than about 9 ins (228 mm) wide.

PLANK CONSTRUCTION:
1. timber construction making use essentially of planks.
2. the use of heavy timbers of plank-like cross-section, though perhaps of greater dimensions, to make a structure, rather than posts of square, or nearly square, cross-section. (d238,239)

PLASTER CLADDING: *see* CLADDING

PLATE: a piece of structural timber laid horizontally and usually rather wider than deep in proportion.

POLE PLATE: a timber member running from foot to foot of the principal rafters or blades of a series of trusses in order to receive the feet of common rafters and so leave space for a gutter. (d250)

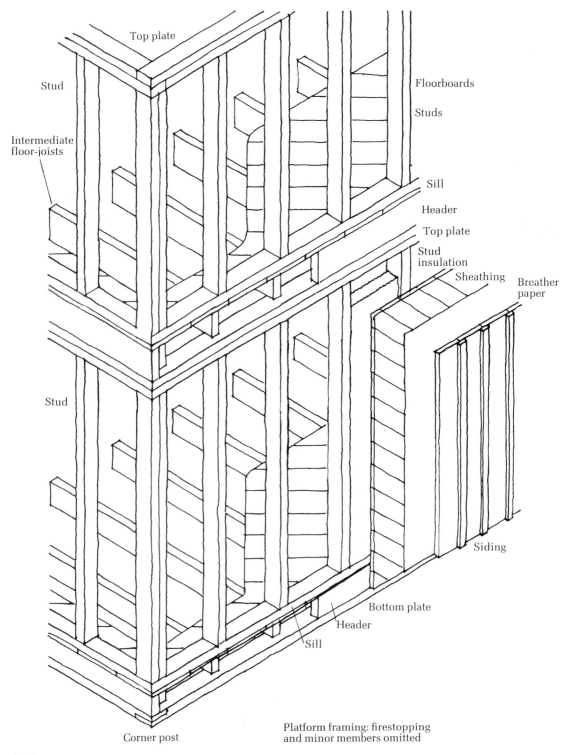

Top plate

Stud

Floorboards

Studs

Intermediate
floor-joists

Sill

Header

Top plate

Stud
insulation

Sheathing

Breather
paper

Stud

Siding

Bottom plate

Header

Sill

Corner post

Platform framing: firestopping
and minor members omitted

d240

160

POST PLATE: a sill from which structural timbers rise. (d251)

ROOF PLATE (TOP PLATE): a timber running horizontally along the top of a timber-framed wall, receiving the tops of studs and carrying the feet of common rafters. It is the equivalent of a wall-plate (d251)

SOLE PLATE: a short member lying horizontally across the top of a solid wall and receiving the foot of a principal rafter or blade. (d252)

WALL-PLATE (PAN, PETTRIL, RAILING PIECE, RAISON, TASSEL): a piece of timber running horizontally along the top of a wall to receive the ends of common rafters. If the wall is a solid wall of brick, clay, flint, stone etc. the wall-plate is carried by the wall and helps spread the load from common rafters uniformly along the top of the wall. If the wall is very thick there might be an inner and outer wall-plate, the outer one receiving the common rafters and the inner one helping with ashlaring. If the wall is timber-framed and consists of posts and studs then the wall-plate receives the heads of these members as well as the feet of the common rafters. If the timber-framed wall is of uniform scantling then the wall-plate again receives studs as well as carrying common rafters. If the timber-framed wall is of normal assembly and carries roof trusses then the wall-plate also helps to locate and receive the tie-beams. Although the terms roof plate or top plate should perhaps be used in timber-frame walling, in practice the term wall-plate is more commonly used. (d253)

PLATED YOKE: see YOKE

PLATFORM FRAMING: a type of lightweight timber-frame construction developed in North America in which sawn studs, usually of 4 ins by 2 ins cross-section (100 mm by 50 mm) run from a timber sill to a top plate to make one storey in height. Floor-joists are carried by the top plate and their ends then carry in turn another sill from which studs rise to another top plate and so on, storey by storey. Triangulation is provided with the aid of diagonal boarding or plywood sheathing nailed to the outside of the studs. The construction depends on nailing rather than jointing.

Platform framing differs from balloon framing in that studs only run through one storey height. This means that it is quite easy to prefabricate sections of wall consisting of sill, studs and top plate. (d240)

PLYWOOD: a board made from thick sheets of wood veneer glued together in threes, fives or other odd numbers, with grain running at right angles to each other in alternate sheets. The veneer is cut by revolving a circular log

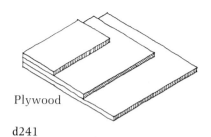

Plywood

d241

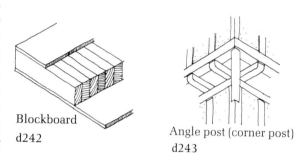

Blockboard
d242

Angle post (corner post)
d243

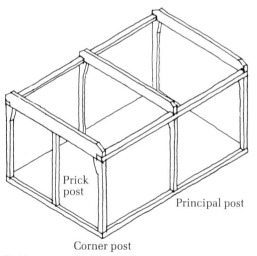

Prick post

Principal post

Corner post

d244

against a plane. Plywood is usually associated with interior finishes but exterior quality plywood is available and used for structural purposes. (d241) BLOCKBOARD has thin strips of wood glued between veneers. (d242)

POLE PLATE: see PLATE.

POST: a long timber member set vertically and intended primarily to carry a load imposed on its top.

AISLE POST (ARCADE POST): the post carrying arcade plate and tie-beam in aisled construction. (d16a)

ANGLE POST: the post at the corner of a building above which a storey jetties in two directions. Often the angle post has its head tilted outwards to help carry a dragon beam as part of the jettied construction. (d243) (**79**)

ARCADE POST: *see* AISLE POST above.

CORNER POST: a post at the corner or at the wing of a building. (d244)

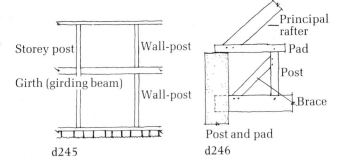

Storey post | Wall-post
Girth (girding beam) | Wall-post

Principal rafter
Pad
Post
Brace
Post and pad

d245   d246

PRICK POST: an intermediate post not carrying the end of a tie-beam and thus usually found in the gable or hip wall of a building. Prick posts are usually only one storey in height and differ from studs in that they give intermediate support to some types of beam. (d244)

PRINCESS POST: *see* QUEEN POST ROOF TRUSS

PRINCIPAL POST (MAIN POST, TEAGLE POST): one of the posts at corner or bay intervals marking one of the main structural divisions of a building. (d244)

STOREY POST: a wall-post which runs through two or more storeys in a multi-storey building. (d245)

WALL-POST: a post which forms part of a wall as distinct from one which is free-standing, as in aisled construction. (d245)

POST AND BEAM CONSTRUCTION:
1. a general term for timber-frame construction incorporating posts, longitudinal beams (wall-plates) and transverse beams (tie-beams).
2. in North America a specific term for a timber-frame construction based on planks spanning between transverse beams each about 8 ft (2.4m) apart and carried on pairs of posts.

POST AND PAD: a form of roof construction in which a principal rafter or blade rises from the end of a short timber pad the other end of which is carried by a short timber post rising from a floor beam, such as a binder. The post has a diagonal brace running from near the end of the

floor beam to the top of the post just beneath the pad. (d246) Post and pad construction was used where a clear central space was required within a roof as for a granary, warehouse or workshop.

POST AND PAN (POST AND PETREL, POST AND BETREL): a term once used in the North of England, and especially in Lancashire, for timber-frame construction in its basic form of posts and wall-plates. Sometimes the word 'pan' is mistakenly assumed to be a term for a panel, but in context it is clear that wall-plate is intended in documentary references.

POST- AND TRUSS-CONSTRUCTION: timber-frame construction based on a series of transverse frames or cross-frames. The frames include roof trusses with tie-beam and posts; there are end-frames at each end of the building and intermediate frames along its length; the frames are spaced and linked by means of wall-plates. *See also* BOX-FRAME CONSTRUCTION. (d36a, b)

PRICK POST: *see* POST

PRINCESS POST: *see* QUEEN POST ROOF TRUSS.

PUNCHEON (PUNCHION, QUARTER, STANCHEON, STANCHION):
1. a short stud such as would be used over a doorway in a timber-framed wall or partition. (d247)
2. a term sometimes used for a stud when it appears to be carrying some load. A puncheon in this sense is lighter than a principal post or a prick post. (d248)
3. the term QUARTER was usually interchangeable with PUNCHEON but was also used specifically for a stud in a timber-framed partition.
4. in North America a floor made out of halved logs. (d249)

Puncheon   Puncheon
                                    Puncheon floor
d247   d248   d249

79. Angle post and jettying, The Chantry, Sudbury, Suffolk

80. Purlins, barn at Avebury, Wilts.

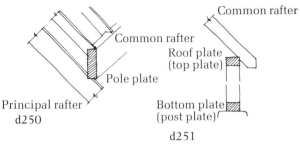

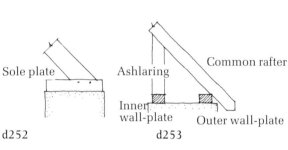

d250

d251

d252

d253

PURLIN: a longitudinal member giving support to the common rafters of a roof and normally set at right angles to the slope of the rafters.

BUTT-PURLIN (TENON PURLIN, TENONED PURLIN): a longitudinal member giving intermediate support to common rafters but tenoned into principal rafters. The upper surface of the butt-purlin is set below the upper surface of the principal rafter to the extent of the depth of the common rafter. Butt-purlins are usually staggered to avoid weakening the principal rafter by having too many mortice holes at the same spot. (d254) (80)

CLASPED PURLIN: a longitudinal member giving some intermediate support or lending stability to common rafters, but simply held or clasped in the upper part of the angle between rafter and collar. In earlier examples the rafter is a principal rafter diminished at the junction with the collar; in later examples, common rafters and a light collar are used throughout. (d256) (81)

81. Clasped purlins, house at Carleton Husthwaite, Yorks. N.R.

165

82. Princess posts, Howard Street Warehouse, Shrewsbury, Salop.

COLLAR PURLIN (COLLAR PLATE): a member running horizontally the length of a roof immediately below the collar rafters of a roof and supported at bay intervals by crown posts. It is believed nowadays that the main purpose of the collar purlin was to provide longitudinal stability to pairs of rafters preventing their collapse through racking. The term COLLAR PLATE is sometimes used and is more accurate in that plates are laid square whereas purlins are laid at right angles to the roof slope. (d257)

COMMON PURLINS (HORIZONTAL RAFTERS): timbers of fairly light cross-section running along the roof to sup-

port a roof covering directly, or by way of boarding nailed to these timbers. The common purlins are supported by trusses, often fairly closely spaced. Common purlins were widely used in North America during the seventeenth and eighteenth centuries to support roof coverings of shingles or long wooden shakes. Though

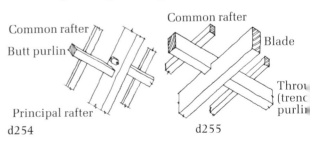

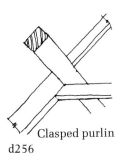

Clasped purlin
d256

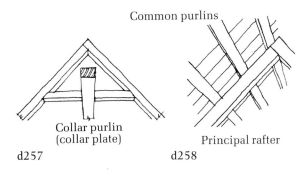

Common purlins

Collar purlin
(collar plate)
d257

Principal rafter
d258

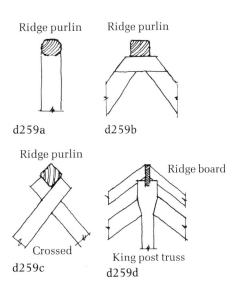

Ridge purlin   Ridge purlin

d259a   d259b

Ridge purlin   Ridge board

Crossed   King post truss
d259c   d259d

sometimes to be found in Scotland, common purlins are rarely seen in England and Wales except in boarded roofs or in roofs covered with corrugated sheeting. (d258)

DORMENT (DORMANT): an obsolete term for a purlin.

RIB: an obsolete Cumbrian term for a purlin.

RIDGE PURLIN (RIDGE BEAM, RIDGE PIECE, ROOF TREE): a longitudinal member receiving the upper ends of common rafters at the apex of a roof. The ridge purlin may be simply a rounded log or, if squared, set horizontally (when strictly it should be called a plate). It may be set diagonally, approximating to the angle of the roof slope. (d259a, b, c)

RIDGE (RIDGE BOARD): a light, plank-like member running from truss to truss and receiving the ends of common rafters, but of much more slender cross-section than a ridge purlin or side purlin. (d259d)

SIDE PURLIN: a longitudinal member giving intermediate support to common rafters and located between wall-plate and ridge level. There may be one, two or more side purlins to each roof slope, depending on the length of the sloping side of the roof and the nature of the roof covering. Side purlins may be fashioned as butt-purlins or as through purlins. (d36b, 65)

SIDE REZOR, SIDE WEVOR: obsolete terms for side purlins.

SLEEPER: cf. DORMENT: another obsolete term for a side purlin.

THROUGH PURLIN (TRENCHED PURLIN, LAID-ON PURLIN): a side purlin carried on the backs of blades in roof trusses. Side purlins may run continuously in one cross-section along the whole length of a roof (subject only to scarf jointing) or may be in sections between trusses. A through purlin may rest on the back surface of a truss blade, located by a cleat, or may be jointed e.g. by cogging to the blade. (d255)

TRUSSED PURLINS: purlins which are in effect deep trussed beams. A timber top boom or longitudinal member carries the common rafters; ceiling joists may be carried by a timber bottom boom. A calculated pattern of vertical and inclined members which may be of timber, wrought iron or both, provides triangulation. Since considerable depth is usually available, trussed purlins can be designed to carry loads over considerable spans—from end wall to end wall in a house, for instance—or between widely spaced trusses. (d260, 261, 262)

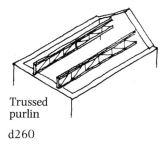

Trussed
purlin
d260

QUARTER: see PUNCHEON and STUD

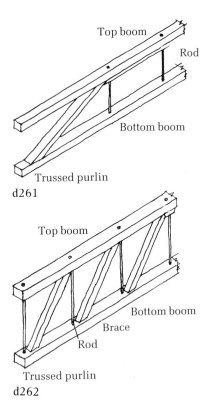

Top boom

Rod

Bottom boom

Trussed purlin

d261

Top boom

Bottom boom

Brace

Rod

Trussed purlin

d262

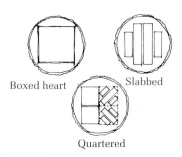

Boxed heart

Slabbed

Quartered

d263

83. Queen post roof truss, Albert Warehouse, Gloucester

QUARTERING: a method of conversion of a log into four segments of timber (d263).

QUEEN POSTS: members rising vertically from a tie-beam to give direct support to side purlins and symmetrically disposed about the centre of a roof. (d264)

QUEEN POST ROOF TRUSS: a roof truss consisting essentially of a tie-beam from which rise two queen posts each carrying a side purlin and receiving the upper part of an inclined blade which rises from one end of a tie-beam. The upper parts of the queen posts are jointed to a heavy horizontal straining beam and a lighter straining sill runs between the feet of the queen posts and rests on top of the central part of the tie-beam. Braces run diagonally between the foot of each queen post and the inclined

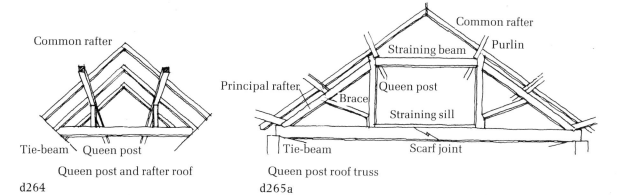

Common rafter

Common rafter
Purlin
Straining beam
Queen post
Principal rafter
Brace
Straining sill
Tie-beam
Tie-beam
Scarf joint

Queen post and rafter roof
d264

Queen post roof truss
d265a

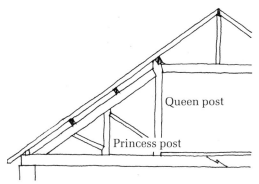

Queen post

Princess post

Queen post roof truss with princess posts
d266

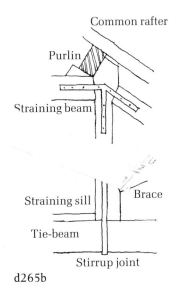

Common rafter

Purlin

Straining beam

Straining sill

Brace

Tie-beam

Stirrup joint
d265b

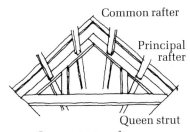

Common rafter
Principal rafter

Queen strut

Queen strut roof truss

d267

blade. Where the span of the truss is considerable, subsidiary posts called PRINCESS POSTS rise at the angle of tie-beam and inclined blades. (d265a, b, 266) (**82**) (**83**)

Because of the spans for which queen post roof trusses are intended, the tie-beam is usually scarfed or fabricated in some way out of several pieces of timber.

QUEEN STRUT: one of a pair of vertical timbers rising from a tie-beam to a collar or a principal rafter and possibly giving indirect support to side purlins. A queen strut does not carry side purlins directly. (d267)

RACKING: the tendency of a set of timbers to collapse lengthways.

RADIOCARBON DATING (CARBON 14 DATING): a method of establishing the felling date of a piece of timber by reference to the radioactive carbon content. The carbon 14 isotope begins to lose its radioactivity as soon as the tree is killed by felling. By testing in a laboratory sections taken from a piece of timber it is possible to measure the decay of radioactivity and thus a probable felling date.

RAFTERS: inclined members following the slope of the roof.

COMMON RAFTERS (CABERS, SPARS): timbers of fairly slight cross-section following the slope of the roof to support a roof covering. Common dimensions for early rafters were 4 ins by 4 ins (100 mm by 100 mm), and 4 ins by 2 ins (100 mm by 50 mm) for later rafters. Spacing of rafters has commonly been between 16 ins (406 mm) and 24 ins (610 mm) apart. Common raf-

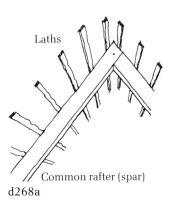

Laths

Common rafter (spar)

d268a

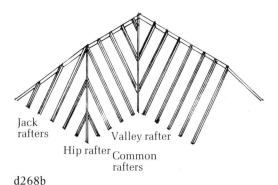

Jack rafters

Valley rafter

Hip rafter

Common rafters

d268b

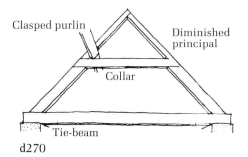

Principal rafter

Butt purlin

Tie-beam

d269

Clasped purlin

Diminished principal

Collar

Tie-beam

d270

In Scotland and the North of England common rafters are still called SPARS while in Scotland they may also be called CABERS.

HIP RAFTER (q.v.) receives JACK RAFTERS at the angle of the inclined planes of a hipped roof. (d268b)

JACK RAFTERS: shortened common rafters running between eaves and hips or between ridge and valley. (d268b)

PRINCIPAL RAFTERS: heavy timber members following the slope of a roof and forming part of a roof truss. R. A. Cordingley suggested that the term should be confined to those instances in which the upper surface of the principal rafter coincided with that of the common rafter; where the principal rafter was set lower he suggested the term blade. This distinction is not made by all other classifiers nor is it made in earlier glossaries. (d269) DIMINISHING PRINCIPALS (REDUCED PRINCIPALS) are principal rafters in which the upper parts above collar level are cut back to the depth of common rafters. (d270) KERB PRINCIPALS (SHORT PRINCIPALS) are principal rafters rising from a tie-beam only as far as a collar. (d271) KNEES OF PRINCIPAL RAFTERS: sometimes the ends of principal rafters were cranked so as to enter the ends of the tie-beams at right

Kerb principal

d271

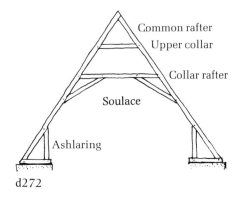

Common rafter

Upper collar

Collar rafter

Soulace

Ashlaring

d272

ters might extend from eaves to ridge in one piece but where supported by purlins they were often divided into shorter lengths running from purlin to purlin. (d267, 268a, b)

angles. These knees were cut out of the solid but then appear similar to those of upper crucks. The technique was most used in the late seventeenth century and was intended to concentrate roof loads on supporting walls

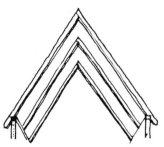

d273   Coupled rafter roof

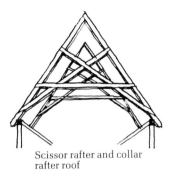

Scissor rafter and collar
rafter roof
d276

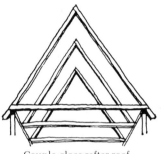

Couple-close rafter roof
d274

d277   Collar rafter roof

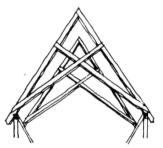

d275   Scissor rafter roof

Queen post and collar rafter roof
d278

as well as improving headroom in roof spaces. *See* HEWN KNEE *under* KNEE.

VALLEY RAFTER (q.v.) receives JACK RAFTERS where two adjacent planes of a roof come together.

RAFTER ROOFS (TRUSSED RAFTERS): these are divided into single and double roofs.

1. RAFTER SINGLE ROOFS: a type of roof consisting entirely of pairs of common rafters with or without ancillary members but not making use of load-bearing purlins.

COLLAR RAFTER ROOF: a roof consisting of pairs of inclined common rafters, each pair linked by a collar. Sometimes there are straight braces (soulaces) between rafter and collar or arch braces rising from post or wall to each collar. There may also be ashlaring. Occasionally, in very steeply pitched roofs there is a second collar above the main collar. (d272, 277)

COUPLE-CLOSE RAFTER ROOF (CLOSE-COUPLE ROOF, SPAN ROOF): a roof consisting entirely of pairs of common rafters but with each pair tied at the base by a tie-beam of similar cross-section to the common rafters. (d274)

COUPLED RAFTER ROOF: a roof consisting entirely of pairs of common rafters though there may be ashlaring at the feet of the rafters if the roof is carried on solid walls. (d273)

SCISSOR RAFTER ROOF: a roof consisting of pairs of common rafters linked by members of similar cross-section running diagonally from the lower third to the upper third of opposite rafters and crossing on the centre line of the roof. There may also be a collar set just below the intersection of the scissor rafters to produce a SCISSOR AND COLLAR RAFTER ROOF. (d275, 276)

2. RAFTER DOUBLE ROOFS: roofs consisting of pairs of common rafters with or without ancillary members but with purlins to give additional support or stability.

CLASPED PURLIN COLLAR RAFTER ROOF: a collar rafter roof with light purlins running the length of the roof and clasped between collar and upper part of common rafters or, with the aid of soulaces or short straight braces, running in the angle between the collar and the lower part of the common rafters.

CROWN POST RAFTER ROOF (CROWN POST AND COLLAR PURLIN ROOF): a roof consisting, like a collar rafter roof, of pairs of common rafters with each pair linked by a collar rafter, but with a longitudinal collar purlin or plate running beneath the collars and supported at bay intervals by crown posts. There may be braces between crown posts, collars and collar purlins as well as between tie-beam and crown post. (d59) (**84**)

QUEEN POST COLLAR RAFTER ROOF: a roof consisting of common rafters linked by collar rafters placed rather high and given intermediate support by side purlins, which are carried on queen posts rising from the tie-beams placed at bay intervals. (d278)

QUEEN POST RAFTER ROOF: a roof consisting of common rafters supported by purlins carried on queen posts. (d264)

RAFTER HOLES (PROBLEM HOLES): holes found sometimes in the sides of common rafters just above the wall-plate. The holes do not, as a rule, pass right through the rafter. They may have been intended for sprockets or intended to help in some way in the alignment of pairs of rafters. (d279)

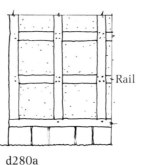

d280a

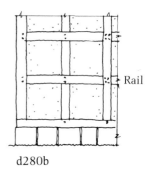

d280b

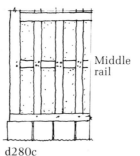

d280c

RAIL: a fairly lightweight horizontal member in a timber-framed wall used with studs to divide bays into panels. Usually rails were jointed into studs but sometimes short studs were jointed into longer rails. (280a, b)

MIDDLE RAIL (MIDRAIL): a rail located about half way up a storey height in a timber-framed wall. (280c)

RAISED CRUCK TRUSS: *see* CRUCK TRUSS

RAISON, RAISING PIECE, RAISING PLATE, REASON, RESON, SIDE-RESON: obsolete terms for a WALL-PLATE *see under* PLATE

RAKING BEAM: the inclined blade forming part of a roof truss and carrying through purlins.

RAKING BEAM TRUSS (TRUSS BLADE TRUSS): a roof truss based on the use of raking beams or blades to carry through purlins.

RAKING STRUTS: inclined braces rising from tie-beams to principal rafter or blades. (d281)

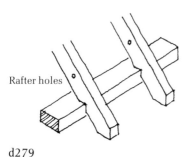

Rafter holes

d279

Raking strut

d281

84. Crown post rafter roof, Church of St. Michael & All Angels, Codford, Essex

REAR, REARING:

1. generally, the process of erecting a timber-framed building.

2. specifically the process of raising into position a part of a building such as a wall frame or a roof truss which has been assembled flat on the ground.

3. more specifically the process of raising into position one of a series of cruck trusses which have been assembled on the ground and are to be reared upright and then man-handled into the selected position. (d8a, b, c, d)

REBATE (RABBET): an open-sided groove cut along the edge of a beam, joist, plank or other piece of timber. Thus early floor-joists were rebated to receive floorboards. (d282)

REVERSED ASSEMBLY: the method of construction in which

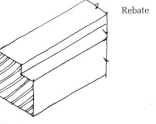

Rebate

d282

the wall-plate is above the tie-beam which in turn is carried by a post. This is in contrast to NORMAL ASSEMBLY (q.v.) in which the wall-plate lies above the post but below the tie-beam. (d212b) (**85**)

RIBBON (BEARER): a term used in North America for a light timber member let into studs and carrying floor-joists in balloon frame construction. (d22, 283)

RIDGE: the apex of a pitched roof. The horizontal line of intersection of the two inclined planes of a pitched roof. (d284)

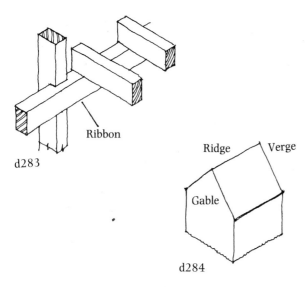

d283

d284

RIDGE PURLIN (RIDGE BEAM, RIDGE PIECE, ROOF TREE): see PURLIN

ROOF PITCH: the slope of a roof; its inclination from the horizontal. Nowadays the pitch is usually quoted in degrees but formerly a rise of so many feet vertically against so many horizontally was quoted.

The roof pitch was determined primarily by the nature of the roofing material. By experience, in a given locality, certain pitches were found to be suitable for certain local materials. Generally, the larger the unit of roof covering the lower the acceptable pitch; thus plain tiles were usually laid at 45° or more whereas large thin Welsh slates could be laid at a pitch as low as $22\frac{1}{2}°$.

ROOF SHAPES: the organisation of the surfaces of a roof as determined by roof covering, constructional methods, plan shape and tradition or fashion.

CATSLIDE ROOF: a roof surface continuing beyond the main eaves or wall-plate level in order to cover an aisle or outshut. (d285c, i)

GABLE (PIGNON, PYNNION): the triangular vertical surface which continues the end wall of a building to the end of a ridge in a gabled roof. (d284)

GABLED ROOF: a ridged double-sloping roof with triangular wall surfaces or gables at each end. DOUBLE-GABLED ROOF has two gables at each end. (d285g, k)

GABLET ROOF: a hipped roof in which small gabled sections are introduced between the upper part of the hipped ends and the ends of the ridge. (d285b)

HALF-HIP AND GABLET ROOF: as HALF-HIPPED but ending at a little gable. (d285f)

HALF-HIPPED ROOF: a ridged double-sloping roof in which the slope is continued round the upper part of the end walls, the lower part being continued upwards as if to form a gable but finishing as a raised eaves. (d285h)

HEEL GABLE: a gable which terminates one of two ranges meeting at right angles. The gable at the end of one roof becomes part of the side wall running below the eaves of another. (d285m)

HIP: the junction between roof slopes on adjacent sides of a hipped-roofed building. (d285a)

HIPPED ROOF: a ridged double-sloping roof which is continued round one or both ends. DOUBLE-HIPPED ROOF has two hips at each end. (d285a, d)

MANSARD ROOF (GAMBREL ROOF): a roof having a steep slope rising from the eaves and a shallower slope reaching the ridge with an angle or kerb between the two. (d285j) (**86**) A HIPPED MANSARD ROOF is a hipped roof with steep slopes from the eaves and shallower slopes reaching the ridge. This is most nearly similar to the type of roof made popular in France during the seventeenth century by the architect Francois Mansard whose name has been taken to describe the roof. (d285e)

M-SHAPED ROOF: a roof with a central longitudinal gutter set at a higher level than the eaves. Normally the ends of such a roof are gabled and so give the characteristic M shape but the same principle can be applied to a hipped roof or a hipped mansard roof. The M-shaped roof is designed to cover a building of deep plan while allowing the roof space to be used. (d285n)

PENT ROOF (PENTISE): a roof of single pitch built against a wall. When roofing an open gallery often called a PENTISE. (d285l, o)

ROOF TRUSS: a set of stout timbers framed together to support a range of common rafters by means of purlins. Roof trusses are positioned along a wall partly according to the

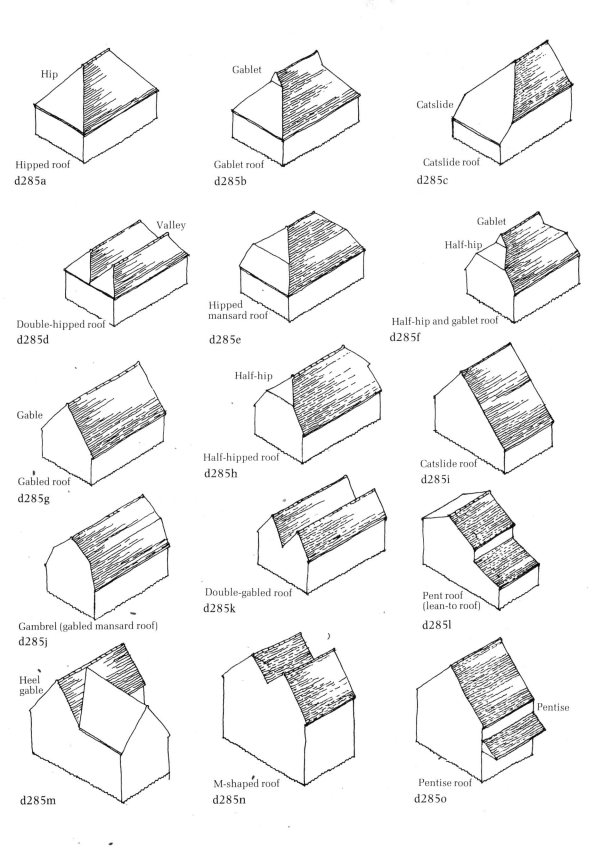

Hip

Hipped roof
d285a

Gablet

Gablet roof
d285b

Catslide

Catslide roof
d285c

Valley

Double-hipped roof
d285d

Hipped mansard roof
d285e

Gablet

Half-hip

Half-hip and gablet roof
d285f

Gable

Gabled roof
d285g

Half-hip

Half-hipped roof
d285h

Catslide roof
d285i

Gambrel (gabled mansard roof)
d285j

Double-gabled roof
d285k

Pent roof (lean-to roof)
d285l

Heel gable

d285m

M-shaped roof
d285n

Pentise

Pentise roof
d285o

placing of supports, partly according to the desired length of unsupported purlins, and partly according to the planning requirements for the spaces to be covered by the roof.

RUNNER: a light timber member, usually running horizontally in a roof, suspended at intervals by a hanger from a blade or rafter and to which ceiling joists are nailed in order to be given intermediate support. (d105)

SADDLE: a short piece of timber carrying a ridge purlin and receiving the upper ends of truss blades or principal rafters. (d286)

SAMSON POST: a term derived from ship construction but applied to buildings to describe timber posts rising from a floor to carry a beam. Samson posts were used, for instance, when an intermediate floor was to be inserted in a tall room or in industrial buildings where intermediate support was to be given to heavy binders or girders. Loads were usually concentrated on to the ends of samson posts with the aid of pillows (q.v.). (d237) (**87**)

SCANTLING:
1. a term used generally for the cross-section of a piece of timber thus a floor-joist might be of 6 ins by 3 ins (151 mm by 76 mm) scantling.
2. a term used more specifically for light timbers with cross-section up to 5 ins (127 mm) square. Thus a timber-frame partition with studs 4 ins by 2 ins (100 mm) by 50 mm) might be described as made up of scantlings.

UNIFORM SCANTLING: the use of lightweight timbers of a single cross-section. Thus a timber wall of balloon frame construction would be of uniform scantling in that every member was of 4 ins by 2 ins (100 by 50 mm) cross-section.

SCARF, SCARFED JOINT, SCARFING: *see* JOINTS

SCARFED CRUCK: *see* CRUCKS

SCOTCHES (HOISTING GROOVES; STAY NOTCHES): shallow tapered recesses found sometimes on the outer surfaces of posts and believed to be intended as a secure anchor for the ends of temporary props used during the erection of a timber-framed building or its repair, especially by under-pinning. (d287a, b, c)

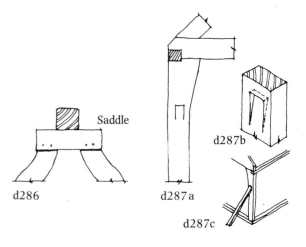

d286   Saddle

d287a

d287b

d287c

SECRET JOINTING: *see* JOINTS

SECRET NAILING: *see* FLOORBOARDS

SHORE: a temporary support to a wall. (d292, 293) (**88**)

SHORT KING POST: *see* KING POST ROOF TRUSS *under* KING POST

85. Reversed assembly, Old Garland Hall, Thelwall, Lancs.

86. Gambrel roof, tide mill, Woodbridge, Suffolk

SIDE PURLIN: *see* PURLIN

SIDE RAISON (SIDE RESON): *see* WALL-PLATE *under* PLATE

SIDE WYVER: obsolete term for a PURLIN

SILE: a North Country term for a CRUCK BLADE; *see under* CRUCKS

SILL (CILL, GROUND PLATE, GROUND SELL, GROUND SILL, SELL, SILL BEAM): the plate at the foot of a timber-framed wall from which rise all the studs (and usually the posts also). The sill generally rests on a foundation wall. (d288)

SINGLE ROOF: *see* RAFTER ROOF

SKEW-PEGGING: the use of pegs set at an angle to help tighten a joint. (d289)

SLAB: the part rounded piece of timber sawn off when a log is converted into squared timber. (d290)

177

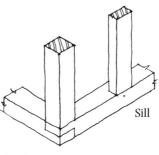

d288

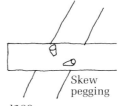

Skew pegging

d289

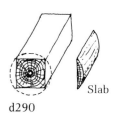

Slab

d290

Sill

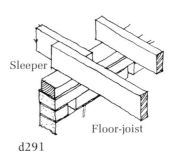

Sleeper

Floor-joist

d291

SLEEPER:
1. a plate laid on a sleeper wall and carrying the joists of the ground floor of a building. (d291)
2. a floor-joist at the ground floor of a building.
3. an obsolete term for a valley rafter (q.v.).
4. an obsolete term for a beam such as a side purlin: *see* DORMENT.

SLING BRACE: *see under* BRACE

SLING BRACE TRUSS (SLING TRUSS): a roof truss in which a truncated tie-beam extending from the end of the principal rafter or blade is received by a sling brace rising from a post to the underside of the principal rafter or blade. Such a truss was used in a garret or granary where head-room was needed within the roof space. (d40)

SLIP TENON: *see* JOINTS

87. Pillow, Herbert Warehouse, Gloucester

88. Raking shores, Kinnoull Lodging, Perth, Scotland

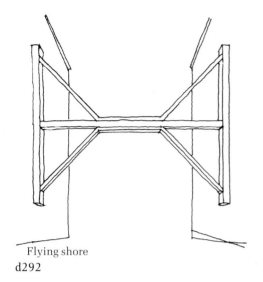

Flying shore
d292

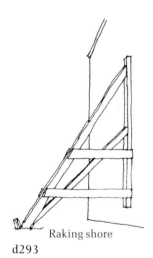

Raking shore
d293

SOFFIT BOARD: *see under* EAVES

SOFTWOOD: *see* TIMBER

SOLE PIECE: a short horizontal member lying across a wall top, normally carried by outer and inner wall-plates and receiving the end of a common rafter and the foot of an ashlar piece in an ashlared rafter roof. (d20)

SOLE PLATE:
1. a type of PLATE receiving the end of a principal rafter or blade and the foot of an ashlar post in an ashlared roof: *see* PLATE.
2. an alternative term for a pad in the sense of a truncated tie-beam lying on top of a post head and receiving the

end of a principal rafter or blade. Often the member is extended inwards and joined to a deep arch brace.

SOLID STRUTTING: *see* STRUTTING

SOULACES: diagonal braces rising from common rafters to collar rafters in a rafter roof. Soulaces may act with ash-laring to help produce a multi-facetted roof left exposed or lined with boarding or plaster. (d272)

SPAN ROOF: an alternative term for a RAFTER ROOF

SPANDREL TIE: a short horizontal or inclined member connecting a post head and an arch brace.

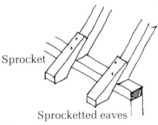

Sprocket

Sprocketted eaves
d294a

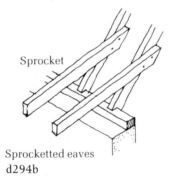

Sprocket

Sprocketted eaves
d294b

SPARS: *see* COMMON RAFTERS

SPERE: a short length of partition projecting into a room to give some draught protection. Most often a spere partition projects inwards from an outer wall as part of a screens passage between opposed doors at one end of a hall open to the roof. Speres projecting inwards from the two outer walls terminate in stout and often decorated spere posts which rise to the tie-beam and so form a spere truss. This usually carries a pair of plates or purlins laid on the flat. The term may also be used for a projection from the end wall of an open hall into the length of the room and giving some draught protection to a dais. At a more humble level the term is applied to the timber partition running just inside the front door of a cottage or farm-

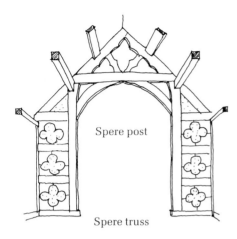

Spere post

Spere truss

d295

house and giving some little draught protection to the principal room. (d295)

SPERE TRUSS: a spere together with posts and tie-beam. (d295) (**89**)

SPROCKETS (CHANTLATES): short pieces of timber projecting from the feet of common rafters to carry the roof covering over the head of a wall and to provide eaves protection. The pitch of the sprocket is usually rather less than that of the main roof. Sprockets are attached either to the tops or the sides of the common rafters. (d294a, b)

SPUR TIE: an alternative term for a CRUCK TIE (q.v. *under* CRUCK SPUR)

STAFF: a lightweight piece of split timber sprung into grooves to provide reinforcement for a daubed wall either directly or with woven wattles. (d213c)

STEMMING: *see* STRUTTING

STIRRUP: *see* FASTENINGS

STOOTHED PARTITION: *see* QUARTERED PARTITION *under* PARTITIONS

STOREY POST: *see* POST

STRAINING BEAM: a stout timber member placed horizontally between the heads of queen posts as part of a queen post roof truss, and serving primarily to prevent the queen posts from collapsing inwards. (d265)

89. Spere truss, Smithills Hall, Bolton, Lancs.

STRAINING PIECE: a fairly light piece of timber introduced beneath the upper horizontal members and between the main vertical members of a trussed partition. (d230, 231)

STRAINING SILL: in a queen post roof truss a fairly light piece of timber added to the top surface of the tie-beam and fitted tightly between the feet of the queen posts to help prevent the feet of the queen posts from moving inwards. The straining sill may also be bolted and keyed to the tie-beam to increase its depth at the centre of the span. Where princess posts have been included in a queen post roof truss of large span, extra straining sills may be introduced between queen posts and princess posts. (d265)

STRAP: *see* STIRRUP *under* FASTENINGS

STRUT: a member intended to act in compression so as to prop or otherwise reduce the unsupported length of a major structural member. Struts may be vertical or inclined, are usually short, and are usually approximately square in cross-section. As a prop, a strut differs from a brace in contributing nothing to triangulation. (d296)

QUEEN STRUTS: struts set on either side of the centre line of a truss. (d296)

V-STRUTS: are inclined struts meeting at the base as in the letter V. They may be found in the upper part of a roof truss rising from a collar to meet the principal rafters or blades. (d296)

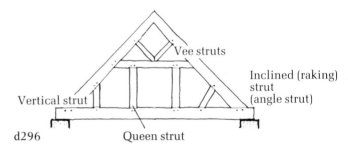

Vee struts

Inclined (raking) strut (angle strut)

Vertical strut

d296   Queen strut

STRUTTING (BRIDGING): a device to increase the rigidity of floor construction. To help prevent tall narrow floor joists from collapsing sideways, or buckling to one side under load, strutting is introduced at intervals usually of about 6 ft (1.8m). The two main types are SOLID STRUTTING consisting of narrow battens wedged in position in a line between the joists, and HERRINGBONE STRUTTING of battens wedged (or more commonly nailed) into position between the top of one joist and the bottom of another running in both directions criss-cross between joists.

CROSS-NOGGING is an alternative term for HERRINGBONE STRUTTING and STEMMING for solid strutting. The term used in North America for strutting is BRIDGING. An obsolete

90. Strutting, Biddles Warehouse, Gloucester

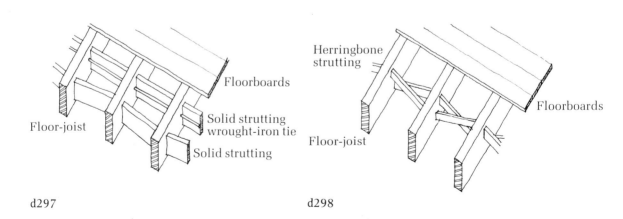

Floorboards

Solid strutting
wrought-iron tie

Floor-joist

Solid strutting

Herringbone
strutting

Floorboards

Floor-joist

d297

d298

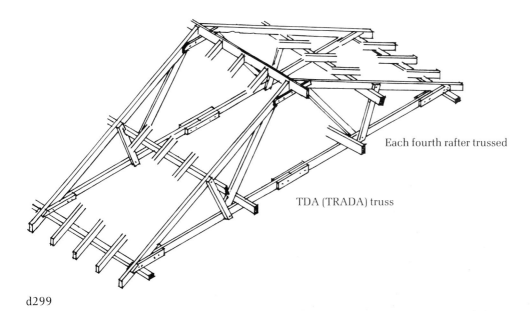

Each fourth rafter trussed

TDA (TRADA) truss

d299

term (which may still survive in Scotland) for a piece of strutting is a DWANG. (d297, 298) (**90**)

STUDS (PUNCHEONS, QUARTERS): lightweight timber members running vertically to help divide a wall or partition into panels. Normally studs are considered as non-structural and used between posts, but in a uniform scantling wall or partition the studs are structural even though each only carries a small proportion of the load. (d213, 215, 227)

STUMP TENON: *see* JOINTS

SUMMER, SUMMER BEAM, SUMMER TREE: *see* BRESSUMMER.

TAIL TRIMMER: *see* TRIMMING JOIST *under* JOISTS.

T.D.A. ROOF TRUSS (T.R.A.D.A. ROOF TRUSS): a type of trussed rafter system devised by the Timber Development Association (later the Timber Research and Development Association). The trussed rafters have joints formed with connectors and are spaced at intervals of about 6 ft (1.8m) along the length of a roof and carry lightweight clasped purlins which in turn help support common rafters. (d299)

TEAZLE TENON: the tenon which rises from the head of a post to engage in a mortice hole in the underside of a tie-beam. At the junction between post, wall-plate and tie-beam in a timber-framed structure of normal assembly there are two tenons, one which runs horizontally to help locate and secure the wall-plate, the other, rising from the

upper and inner part of the post runs laterally to help locate and secure the tie-beam. This latter tenon is the teazle tenon. Usually, it runs on the centre-line of the post head but in many structures, built from the early seventeenth century on, the teazle tenon was moved off-centre. (d212a, b)

TEMPLATE
1. a piece of timber receiving the end of a beam, principal rafter or blade and spreading its load along a wall. The end of the load-bearing member might be tenoned into the template. (d101) (**5**)
2. a wooden model giving the shape to be followed by a carpenter working a piece of timber.

THROUGH PURLIN (TRENCHED PURLIN): *see* PURLIN

TIE-BEAM: *see* BEAM

TIE-BEAM TRUSS: a roof truss consisting essentially of two principal rafters or blades and a tie-beam. There may also be other straight or inclined struts or braces and collars as subsidiary members. (d269)

TIMBER: Timber trees are those which are still growing but are likely to be useful for structural purposes in building or engineering works; such trees when felled and converted become timber. Although a general term, 'wood' is also more precisely used for the material to be worked by joiners and cabinet makers into windows, doors, fittings and furniture in connection with buildings.

The two principal varieties of timber are hardwood and softwood.

HARDWOOD: a conventional term for the timber of broad-leaved deciduous trees such as oak, beech, sycamore, elm etc. Such timber is not necessarily 'hard' for working purposes, though usually it is relatively hard.

SOFTWOOD: a conventional term for the timber of coniferous trees such as fir, pine, spruce etc. Such timber is usually softer and more easily worked than hardwood, but not necessarily so.

TOP PLATE: *see* ROOF PLATE *under* PLATE

TRAPPED PURLIN ROOF: an alternative term for CLASPED PURLIN ROOF (*see under* PURLIN)

TREE-RING DATING: *see* DENDROCHRONOLOGY

TRENAILS: a term sometimes used for PEGS especially when of large diameter.

TRENCHED PURLIN: *see* THROUGH PURLIN *under* PURLIN

TRIANGULATION: by the principle of triangulation a joint which would otherwise be able to move could be made rigid. Normally triangulation involves the use of a member passing diagonally across or between the two members which are to be secured. Braces act as triangulating members.

TRIMMED JOIST, TRIMMING JOIST, TRIMMER JOIST: *see* JOISTS

TRIMMING: the process of arranging and supporting joints around an opening either within a floor, as for a staircase, or at one side of a floor, as for a fireplace or flue. The term is also used for the corresponding process in a roof involving ceiling joists or common rafters. (d185,186)

TRUSS: a frame consisting of several pieces of timber, jointed and triangulated in order to retain its shape under load. When such members are put together in this way they are said to be TRUSSED.

A ROOF TRUSS is a truss made up of relatively stout timbers and placed with others at bay intervals along a building to support purlins, which in turn support common rafters. A CLOSED TRUSS is one incorporating a tie-beam. An OPEN TRUSS is one in which the tie-beam, considered inconvenient or unsightly has been omitted and the necessary triangulation achieved in some other way. (d19a, b, 36b)

TRUSSED BEAM: a beam which consists of several members (rather than a single piece of timber) jointed and triangulated so as to be load-bearing. (d32)

TRUSSED PARTITION: *see* PARTITIONS

TRUSSED PURLIN: *see under* PURLIN

TRUSSED RAFTERS: these are assemblies of common rafters and other members, such as collars and scissor braces, which are jointed and triangulated and so trussed in order to be load-bearing. (d274, 275, 276)

UNDERPINNING:
1. the process of building foundation walls of stone or brick up to the underside of a timber sill when a timber-framed structure has been assembled on temporary cradles.
2. the process of replacing parts of walls in the course of repair after decay, for instance, and involving temporary support of a timber-framed structure with the aid of needles, props, shores and scotch holes. (d300a, b)

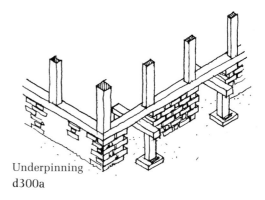

Underpinning
d300a

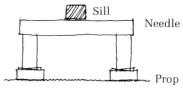

Sill
Needle
Prop

d300b

UPPER CRUCKS: *see* CRUCKS

UPPER CRUCK TRUSS: *see* CRUCK TRUSS

UPPER KING POST: *see* KING TIE *under* KING POST

UPSTAND: in a crude type of joint between a post, wall-plate and tie-beam, the wall-plate is held against an inside or outside UPSTAND rather than a jowl. (d305a, b)

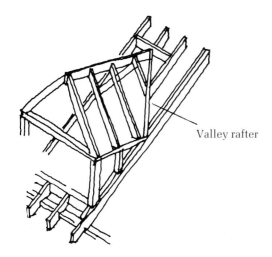

Valley rafter

d301

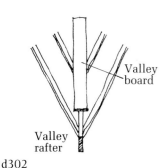

Valley board

Valley rafter

d302

VALLEY BOARD: a wide board acting as the basis of a valley gutter at the internal angle of two meeting roofs. (d302)

VALLEY RAFTER: a deep timber member following the internal angle at the junction of the two slopes of a roof of two parts, as where a wing or dormer meets a main roof. The valley rafter receives the ends of shortened or jack rafters. It is the inward counterpart of an outward-facing hip rafter. (d268, 301)

V-STRUTS: *see* STRUTS

VERGE: the intersection of a gable wall and roof covering in a sloping roof. Where the roof covering extends considerably over the wall there is often a barge board at the verge. (d284)

WALL PIECE: a timber member rising vertically from a corbel against the inner face of a brick or stone wall. Where the wall piece is used against a gable wall it is usually intended to carry the end of a purlin, perhaps with the assistance of an angle brace. Where the wall piece is used against a side wall it rises so as to receive one end of a roof truss at its top and an arch brace at its bottom. (d303)

WALL-PLATE: *see under* PLATE

WANEY TIMBER: where a piece of timber cannot be brought to sharp arrises at the junction of the various surfaces because the edges are near to the rounded outline of the log from which it has been converted, the piece is said to be WANEY or to have WANEY EDGES. (d304)

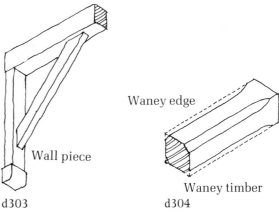

Waney edge

Wall piece

Waney timber

d303   d304

WATTLE AND DAUB (RAD AND DAB): a panel infill consisting of wattles woven around staves and daubed both sides with a clay-like mixture. The wattles consist either of split oak laths or of withies of hazel or willow. The staves are usually of cleft oak, pointed at one end to fit into an auger hole, tapered at the other end so as to be sprung into a groove. The daub is of earth or clay mixed with manure and road sweepings and moistened sufficiently to be daubed on the wattlework. Usually the daub was finished inside and out with a coat of lime plaster or of very fine earth daubing. (d213, 214, 215, 216, 217)

WEDGE: a tapered piece of wood used to tighten a joint or adjust a support. (d153)
FOLDING WEDGES are used for instance, to adjust shoring or underpinning. (d162). **FOXTAIL WEDGING** is used to tighten a tenon against the bottom of a mortice. (d152)

WIND BEAM (WIND BAND):
1. an obsolete term for a collar.
2. a sort of passing brace, running diagonally under the common rafters of a roof from the foot of one truss to the head of another, to guard against longitudinal collapse.

WIND BRACE: *see* BRACE

91. Curved yoke, barn at Corrimony, Inverness-shire, Scotland

WRIGHT: *see* CARPENTER

WYVER: *see* SIDE PURLIN *under* PURLIN

YOKE: a piece of timber joining together the upper ends of a pair of blades or principal rafters. (**91**)

COLLAR YOKE: a yoke which fills a triangular space at the top of a pair of blades or principal rafters while joining them. (d306a)

LINK YOKE: a short horizontal member running rather like a high-set collar between the upper ends of two pairs of blades or principal rafters. (d306b)

PLATED YOKE: a member running like a fish-plate across the upper faces of two pairs of blades or principal rafters. (d306c)

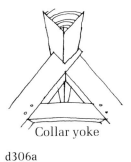

Collar yoke

d306a

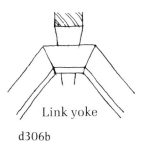

Link yoke

d306b

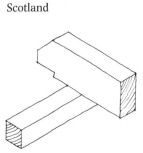

Upstand inside

d305a

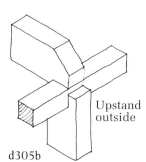

Upstand outside

d305b

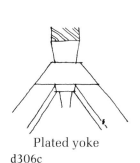

Plated yoke

d306c

Part Three:
Chronological Survey
of Timber Building

In Part Three the photographs illustrate the use of timber in building in Britain from Saxon times until the nineteenth century. The examples are arranged in chronological sequence; some are of buildings bearing dates, the others have been approximately dated. Many show work of more than one period and most have been the subject of repair or restoration. Generally an exterior view has been chosen though much timber construction of great interest is hidden behind the most unpromising of external walls. Although examples have been sought from all periods and many parts of Britain, some periods and counties are exceptionally rich and so have been given an appropriate share of the total.

92. Church of St. Andrew, Greensted-juxta-Ongar, Essex
The whole church was originally built with earth-fast walls of split logs with loose tongues between. Later the bottom parts of the logs were cut away and a sill and foundation wall inserted. Some of the timbers may date to the ninth century.

93. Barn, Grange Farm, Coggeshall, Essex
This barn, at present under restoration, has been dated by Hewett as from about 1140 but with a major rebuild near the end of the fourteenth century.

94. Barley Barn, Cressing Temple, Essex—interior
The interior has been much altered. Originally there were passing braces linking the members on each truss and passing as scissors near the apex. The present roof has a collar purlin braced longitudinally to crown posts which in turn have long slightly curved braces rising from the tie-beams.

95. Wheat Barn, Cressing Temple, Essex—exterior
The Wheat Barn at Cressing Temple has been dated by Hewett to about 1255. The gablet roof falls to low eaves above the close-studded walls of the aisles which surround the main structure.

96. Wheat Barn, Cressing Temple, Essex—interior
The main trusses have been altered but they do show the clasping of a purlin between collar and principal rafter and the partial passing brace running between a scissors crossing near the apex and a joint in the arcade post.

97. Baguley Hall, Manchester
Baguley Hall is notable for the size of the timber members and their flat plank-like shape. Here the heavy spere truss includes deep arch-braces rising to a deeply cranked tie-beam. The members in the partition wall have cusped braces.

98. Stokesay Castle, Shropshire—north tower
At one time thought to be a later modification, the timber-framed projecting storey is now believed to be contemporary with the main stone building of c. 1285–1305.

99. Lower Brockhampton House, Brockhampton-by-Bromyard, Herefordshire
The main building consists of a hall and cross-wing of the late fourteenth century having a spere truss inside the entrance cross-passage and cruck truss in the hall. The walling has square panels except that the upper gabled storey of the cross-wing has tall panels with midrails. The timber-framed gatehouse of a century later has an upper floor jettied all round and carried on angle posts with moulded capitals. The bargeboards have vine trail decoration.

100. Westminster Hall, London
The magnificent hammer-beam roof has been ascribed by Harvey to the period 1394–1400. It replaced an earlier roof structure probably aisled but of unknown form. Although basically of conventional single hammer-beam type the roof structure also makes use of deep arch braces rising from the feet of the wall pieces to the collar.

101. Yardhurst, near Great Chart, Kent—detail
The walls are close-studded with plaster panels little wider than the studs between them. The projecting joists laid flat and carrying the sill of the upper storey are clearly visible.

102. Yardhurst, near Great Chart, Kent
This Wealden house has a hall two bays wide flanked on each side by a bay with jettied upper storey. The wall-plate at the eaves above the hall is in the form of a flying bressummer carried by curved braces. Most Wealden houses are of fourteenth or early fifteenth century date.

103. St. William's College, York

The college was completed in 1466–7 and housed chantry priests of York. It was arranged around a courtyard with an outer face towards the Minster. The ground floor is of magnesian limestone and above rises the timber-framed superstructure. The building was restored in 1910 when the timber oriel windows were added, having been based on the design of an original window in the courtyard.

104. The Ark, Tadcaster, Yorkshire
A hall and cross-wing house. The hall has been rebuilt recently but the cross-wing has been dated by E. A. Gee to the late fifteenth century.

105. Houses and shops in Church Street, Tewkesbury, Gloucestershire
Recently restored, this terrace has been dated by S. R. Jones to the late fifteenth century. Each unit consists of a small shop, and hall on the ground floor and a solar over the shop on the first floor, jettied out to protect the original shopfront. The cambered tie-beam and the slightly curved braces from posts to tie-beam or wall-plate may clearly be seen.

106. Rufford Hall, Rufford, Lancashire
This late example of the medieval hall and cross-wings structure displays an ornately carved spere truss with a free-standing screen between. The spere posts are basically octagonal on plan but heavily moulded at the angles. The wall behind the passage has square panels with plaster quatrefoils sunk into the timber infill. Although restored several times the main house is believed to date from about 1480.

**107. Tudor House,
Newport, Essex**
The wing on the left is
assumed to date from the
fifteenth century and the
main body on the right was
built at various times in the
sixteenth century. The
upper floor is jettied and
curious wavy braces run
between studs and posts.

108. Church of St. Peter, Melverley, Shropshire
This timber-framed church is of early sixteenth-century or late fifteenth-century date though it was restored in 1878. It is quite plain inside and out though the tall plastered panels between very closely spaced studs give quite an extravagant appearance to the exterior.

109. Paycocke's House, Coggeshall, Essex
Built about 1500 this is a superb example of timber framing and decorative woodwork. The whole of the upper part of the front wall is jettied and a fascia board with vine leaf decoration covers the ends of the projecting floor-joists. The tall narrow panels have a brick-nogging infill.

**110. Abbots House,
Butchers Row, Shrewsbury,
Shropshire**

This is a fine example of the
surviving timber-framed
buildings in the heart of
Shrewsbury. Probably of
early sixteenth-century date
the ground floor has
shopfronts of late medieval
type; the first floor is jettied
in two directions and has
tall narrow panels with very
light rails (possibly inserted
at the time the wattle and
daub was replaced by brick
nogging); the second floor,
also jettied in two directions,
has square panelling. There
are decorated angle posts at
ground floor and first floor
corners.

111. Pitchford Hall, Pitchford, Shropshire

A large timber-framed house built on an E-shaped plan between about 1560 and 1570, Pitchford Hall is one of the best surviving examples of its type. The square panels of the timber-framed walls are divided into herringbone and diamond shapes; walls are jettied on all sides; jettied gables project over decorated coving; where panels are not so ornately divided they have closely spaced studs and a midrail.

112. Churche's Mansion, Nantwich, Cheshire

Dated 1577, Churche's Mansion is located outside the main part of Nantwich and so escaped the fire of 1583. It is multi-storey throughout and the vestigial hall is squeezed between the projecting porch and the bay windows. First floor and gables are boldly jettied on two sides with coves beneath the gables and decorated fascias and covings beneath the upper floor. Most of the square panels are divided by curved studs, some giving the suggestion of a herringbone effect, others giving an impression of a pointed diamond effect. On the ground floor the closely spaced studs are divided by rails to give three panels to the height of the storey.

113. Staple Inn, High Holborn, London
One of the very few substantial fragments surviving from what was once a timber-framed city, Staple Inn has a hall of 1580, a residential range of 1585 and the range of 1586 which fronts on to High Holborn. There was a major restoration in 1930 and there have been subsequent repairs but the general impression of the late sixteenth century has been maintained. All the floors are jettied but the wall surface of the third floor continues up into the gables and incorporates the long window range which lights the attics.

114. Meeting Lane, Alcester, Warwickshire
The small town of Alcester contains many examples of timber framing especially of the sixteenth and seventeenth centuries. This illustration shows the jettied gable and square framing typical of the district.

115. Crown Hotel, High Street, Nantwich, Cheshire
In 1583 the centre of Nantwich was destroyed by the sort of disastrous fire which occurred at one time or another in most of our ancient towns and cities. However, Nantwich was quickly rebuilt and the Crown Hotel is a good surviving example of the buildings of quality which replaced what had been destroyed. Each floor is jettied with the aid of decorated brackets and the top of the second floor takes the form of a continuous window lighting a gallery. Alongside is another timber-framed building but this has a gable projecting over its narrow frontage towards the street.

116. Church House, Church Lane, Ledbury, Herefordshire
Believed to have been built about 1600, Church House has the familiar combination of square panels on the end of the building and more extravagant use of timber in narrow panels with a midrail on the front. Each floor is jettied and the top floor rises into gables.

117. The Vicarage, Berriew, Montgomeryshire, Wales
This fine timber-framed house displays the date 1616 on its boldly-projecting porch. The main front wall has fairly widely-spaced studs with midrails giving narrow oblong panels. There are long straight braces rising from the sills to the posts. The porch with its jettied walls has small square panels divided into quatrefoils.

118. Gatehouse at Stokesay Castle, Shropshire
Probably built between about 1620 and 1625, the gatehouse has a tall timber-framed ground floor raised quite high on a stone base and with tall panels, midrails and straight braces. The angle posts flare out into bracket-like terminations which carry the bold projection of the jettied first floor. Here the square panels are converted into diamond shapes with diagonal members across each corner. The dormer gable has barge boards and small panels divided by curved members into quadrant and pointed diamond shapes.

119. Tower Hill House, Bromyard, Herefordshire
This square-set house rising from a stone plinth on a sloping site was built in 1630. Both visible elevations have studs with midrails but fairly widely spaced. Jetties have been abandoned but there is a sort of cornice beneath the diamond-shaped panelling at the attic level. The long straight braces rise from the sills to the posts almost at storey height.

120. Market Hall, Ledbury, Herefordshire
This is typical of the many market buildings which once spread over the whole country. There is an open ground floor and a first floor hall carried on heavy timber posts. The gables are divided into square panels while the first floor has herringbone panelling. This market hall was started in 1617 and completed shortly after 1655.

217

121. Upper Bryn, Llanllwchaiarn, Montgomeryshire, Wales
Two timber-framed buildings are set at right angles to each other. Both have square panels, two to each storey, though the panels in fact vary slightly in width and so in proportion. The gabled cross-wing of the longer building bears a restored inscription NOT WE FROM KINGS BUT KINGS FROM US (the motto of the Mortimer family) and the date 1660.

122. Bishop Lloyd's House, Watergate Street, Chester, Cheshire
Like so many of the timber-framed buildings in Chester, a structure basically of the early seventeenth century was thoroughly restored in 1899. The whole of the wall fronting Watergate Street is heavily and ornately carved; at Rows level the superstructure is carried on twisted and bracketed posts; the gables are carried on two sets of brackets at the level of the attic floor.

123. Plas yn Pentre, Aberhafesp, Montgomeryshire, Wales
Peter Smith has shown that this house was built about 1500 as a four-bay cruck-trussed building, the two centre bays being the hall and the two flanking bays containing the subsidiary rooms. A century later an intermediate floor was inserted into the hall, the eaves was raised to its present level (as indicated by the square panels with studs not coinciding with those beneath) and dormer windows were provided to light the newly formed chambers. In 1706 the front wing or porch was added.

124. Farm building, Mottram St. Andrew, Cheshire

The building has the date July 22, 1708 carved on the door lintel. The low-pitched roof is carried on purlins supported directly on blades which are set on the tie-beam well away from the junction with post and wall-plate, a detail common in this part of Cheshire.

125. Cottage at Holmes Chapel, Cheshire
This cottage has a timber frame of square panels raised high on a brick plinth. The brick nogging is almost certainly original; one panel includes a stone plaque bearing the date 1764.

126. Cottages in Ladywell Street, Newtown Montgomeryshire, Wales
In this building of about 1805, the frame of sawn timber is raised on a brick plinth and includes lightweight studs, rails and braces intended for the brick-nogging infill.

127. Warehouse, Howard Street, Shrewsbury, Shropshire
Also known as the Buttermarket, the Howard Street Warehouse was built in 1835 as a market hall for the sale of dairy produce and was located at the terminus of the Shropshire Union Canal. Later it was converted into a railway goods warehouse. There is a semi-basement of brickwork with groined vaults; above rise tall slender cast-iron columns which in turn carry the roof trusses which are heavily strapped with wrought iron.

128. Union Mill, Cranbrook, Kent
Built in 1814, this smock mill consists of a timber-framed boat-shaped cap carrying the sails and revolving on top of an octagonal timber-framed smock which in turn rises from a tall brick base. The whole timber-framed superstructure is clad in narrow feather-edged weather-boarding.

Notes and References

Reference to the works listed in the bibliography is by the surname of the author and the date of publication.

Introduction

1. The definitions of carpentry are from the Architectural Publication Society's *Dictionary* and Nicholson, 1819.

Part One: Pre-framing

2. W. H. Pearsall (as revised for the 2nd. ed. by W. Pennington) shows from pollen analysis that in the Lake District of post-glacial times oak and elm spread thickly up to an altitude of at least 2500 feet above sea level.

3. Carson and others, 1981, pp. 140–1 deals with the stages of timber building by settlers in North America.

4. Ramm, McDowall, Mercer, 1970, p. 67 is one source of the quotation showing the survival of some form of log construction in sixteenth-century England. It quotes from a 1541 survey that in North Tynedale there were then 'very strong houses whereof for the most part the utter sydes or walls be made of great sware oak trees, strong bound and joyned together with tenons of the same so thyche mortressed that yt will be very hard without greatt force to break or cast downe any of the houses . . .'.

5. Charles, 1981, and Rahtz and others, 1982, deal with the transition from earth-fast walling to framing. The matter is also considered in P. J. Drury (ed.) *B.A.R.*, British Series, 110, 1982.

Operations

6. For cultivation of timber for building see Rackham, 1972 and 1976, but an alternative view has recently been put forward by Currie, 1983.

7. Barton, 1979, quotes the 1543 Act 'for the preservation of woodlands', extended in 1570, providing on each acre of coppice wood that after felling, 12 oak standards (or failing oak, elm, ash or beech) should be left. It is believed that the Act was widely disregarded.

8. Tredgold, 1820 etc., recommended that oaks should be felled when about 100 years old (or at least when between 60 and 200 years old); ash, larch and elm when between 50 and 100 years old; spruce and Scots fir when between 70 and 100 years old. In too old a tree the heartwood begins to decay.

9. Harris, 1974, comments on the use of poplar as timber for building.

10. Scots fir and spruce had been planted by 1769 to find a home in the Lake District according to J. D. Marshall, *Furness and the Industrial Revolution*, 1958, p. 60.

11. Salzman, 1952, p. 248 quotes the use of deal in the thirteenth century.

12. On the same page, Salzman, 1952 quotes from the diary of Henry Best, a Yorkshire farmer who in 1641 'went to Hull for deal boards, seasoned, will not draw nails in drying, lighter and cheaper to carry than unseasoned wood'.

13. Neve, 1726, explains that by statute felling was to be performed about the end of April because of the need for bark to be supplied to the tanners, though winter felling was better for the carpenter. The *Dictionary* of the Architectural Publication Society indicates that 'the value of oak bark is so great (in case of small meetings often worth as much as the timber) that . . . the tree is almost always felled in spring when the presence of sap between the wood and the bark renders it easy to strip the former from the latter'. Tyson, 1979 and 1980, has detailed accounts of stripping bark for tanning in Cumbria in the second half of the seventeenth century.

14. Gunther, 1928, p. 236 gives Sir Roger Pratt's calculations.

15. Sir Roger Pratt considered 'the carriage of timber very chargeable and troublesome'. At the rebuilding of Rose Castle in Cumberland in 1669 the accounts show that it cost about 2s. 6d. to fell each tree but an average of 30s. to 'lead' it to the building site (Tyson, 1962, p. 61).

16. For the detailed survey of timber in a Suffolk farmhouse see Rackham, 1972.

17. Rackham, 1972, p. 4 suggests a gradual transition beginning in the fourteenth century from the use of whole tree trunks for major members to sawing lengthwise into two halves. The further transition to sawing into more than two portions became widespread in the late sixteenth century.

18. For tools and their uses see, for example, the fifteenth-century 'Debate of the Carpenters' Tools' in Innocent, 1916, pp. 95–97.

19. See Rackham, Blair, Munby, 1978 for comments on the extent of numbering in carpentry of c. 1270; the Royal Commission on Historical Monuments comments on the early use of numbering on roofing members in Bedern Hall, York (*York* Vol. 5) in the fourteenth century; Moody, 1973, refers to a particularly fine set of marks on ceiling joists of the fifteenth century.

20. Among some clear descriptions and illustrations of numbering see McCann, 1980, Ryder, 1979, Harris, 1978, Charles, 1974, Harding, 1976, Castle, 1977.

21. McCann, 1980, pp. 24–5 deals with carpenters' marks.

22. Moodey, 1973, p. 108 is the source of the allusion to Edward IV.

Jointing

23. The jointing terms used here and in the Glossary are the traditional terms extended by those published by Hewett in his many books and articles, and also by Rigold, 1967, Charles, 1967 etc. and Harris, 1978.

Cruck construction

24. Alcock (ed.), 1981, p. 6 shows 3054 examples of true crucks plotted on a map of England and Wales. In addition the map on p. 72 shows 806 examples of jointed crucks in south-west England to which must be added the 97 examples of jointed crucks in Wales shown on p. 76. The totals of both true and jointed crucks are constantly increasing as fieldwork continues but no change in overall distribution seems to be taking place.

25. An interesting specification for upper crucks is quoted by Tyson 'Some traditional building in the Troutbeck Valley', *Transactions of the Cumberland and Westmorland Antiquarian and Archaeological Society* Vol. 82, 1982, p. 169 in reference to a farmhouse built about 1760: 'the Principals to be Crook footed so that they are a yard from the Floor to the Sparr in the Garrets'.

26. For rearing of cruck trusses see Charles, 1967, and Harris, 1978.

27. Stone cladding of a timber wall to a cruck-trussed building is illustrated on p. 117 in Innocent, 1916.

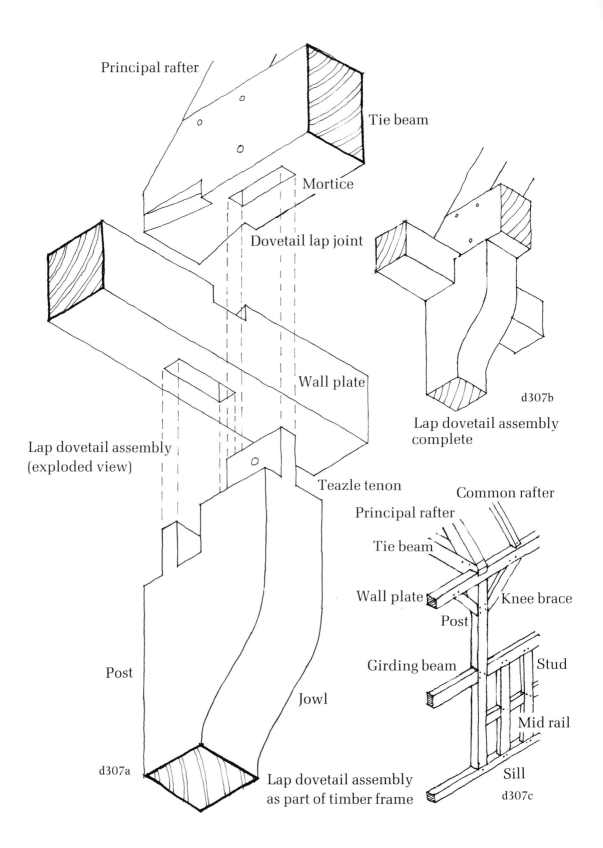

Principal rafter

Tie beam

Mortice

Dovetail lap joint

Wall plate

Lap dovetail assembly
(exploded view)

Teazle tenon

Post

Jowl

d307a

Lap dovetail assembly
as part of timber frame

d307b

Lap dovetail assembly
complete

Common rafter

Principal rafter

Tie beam

Wall plate

Post

Knee brace

Girding beam

Stud

Mid rail

Sill

d307c

Timber-frame walling

28. For definition, use and distribution of the interrupted sill construction see J. T. Smith, 1966, pp. 153–6.

29. For tension in bracing see J. T. Smith, 1966, p. 146 and also Rigold, 1967, Harding, 1976, Martin and Martin, 1974, etc.

30. The mathematical tile dated 1724, discovered by Miss Joan Harding at Westcott, Surrey, is illustrated in R. W. Brunskill and A. Clifton-Taylor, *English Brickwork*, 1977, p. 75.

31. Other examples of farm buildings with jettied upper floors include the Granary, Hyde, Stoke Bliss, Worcs., and the Stable, Colville Hall, White Roding, Essex.

Roofing

32. The distinction between use of butt purlins and through purlins is made in Cordingley, 1961, pp. 74–82.

33. Yeomans, 1981, pp. 9–10 deals with the simultaneous use of the two types of king post truss.

Wide span roofing

34. Rigold, 1967, p. 6 suggested 'shore', as a contemporary term for a member passing from a sole plate to an arcade plate but the term has not found favour.

Part Two: Glossary

35. Hatching on the cross-sections is conventional; it does not necessarily indicate actual growth rings in the timber.

36. The assembly at the head of a post in a timber-framed building is here illustrated in detail. The various constituents appear under separate headings in the glossary but here they have been put together. One diagram gives an exploded view of the assembly, showing the various joints which link post, wall-plate and tie-beam in normal assembly. One can see the importance of the dovetail lap joint in preventing movement and the way the teazle tenon is placed off-centre to help disperse the various cuts in the timber at this point. A second diagram shows the assembly complete and as usually seen from below in a timber-framed building. The third diagram shows the assembly playing its part in the complete system of framing. (d307 a,b,c)

Part Three: Chronological Survey

37. This book is not a constructional text book and no attempt has been made to describe or illustrate the repair and restoration of timber-framed buildings such as are illustrated in this survey. Such matters have been covered as parts of books on the general conservation of buildings and by way of articles in the technical journals. Since this book was written, however, a major work on the subject has been published: F. W. B. Charles' *Conservation of Timber Buildings* has all the authority of one of the most experienced architects practising in this field in Britain today.

Bibliography

S. O. Addy, *The Evolution of the English House*, 1898 (2nd ed. 1905), revised ed. 1933, re-issued 1975

N. W. Alcock and M. W. Barley, 'Medieval roofs with base-crucks and short principals', *Antiquaries Journal*, Vol. LII, 1972, pp. 132–168

N. W. Alcock, 'Warwickshire timber-framed houses, a draft and a contract', *Post-Medieval Archaeology*, Vol. 9, 1975, pp. 212–215

N. W. Alcock, 'What is a gavelfork?', *Vernacular Architecture*, Vol. 8, 1977, p. 830

N. W. Alcock, *Cruck Construction: an Introduction and Catalogue* (CBA Research Reports No. 42), 1981

L. O. Anderson and O. C. Heys, *Wood-frame House Construction*, U.S. Department of Agriculture, 1955

Anon., *Longman's Advanced Building Construction*, 1893

Architectural Publication Society, *Dictionary*, 1851–52 to 1865

T. D. Atkinson, *Local Style in English Architecture*, 1947

M. G. L. Baillie, 'Dendrochronology as a tool for the dating of vernacular buildings', *Vernacular Architecture*, Vol. 7, 1976, pp. 3–10

P. R. Barton, 'Woodland management in the late seventeenth century', *Hertfordshire Archaeology*, Vol. 7, 1979, pp. 181–200

R. Berger (ed.), *Scientific Methods in Medieval Archaeology*, Los Angeles, 1970

J. Bowyer (ed.), *Handbook of Building Crafts in Conservation*, 1981

F. E. Brown, 'Aisled timber barns in East Kent', *Vernacular Architecture*, Vol. 7, 1976, pp. 36–40

Cary Carson and others, 'Impermanent architecture in the southern American colonies', *Winterthur Portfolio*, Vol. 16, Nos. 2/3, 1981

S. A. Castle, 'A late-medieval timber-framed building in Watford', *Hertfordshire Archaeology*, Vol. 5, 1977

F. W. B. Charles, *Medieval Cruck-building and its Derivatives*, 1967

F. W. B. Charles, 'Timber-framed houses of Spon Street, Coventry', *Transactions of the Birmingham and Warwickshire Archaeological Society*, Vol. 86, 1974, pp. 113–31

F. W. B. Charles, 'Scotches, lever sockets and rafter holes', *Vernacular Architecture*, Vol. 5, 1974, pp. 21–24

F. W. B. Charles, 'The timber-frame tradition and its preservation', *Transactions of the Association for Studies in the Conservation of Historic Buildings*, Vol. 3, 1978, pp. 5–28

F. W. B. Charles, 'Post construction and the rafter roof', *Vernacular Architecture*, Vol. 12, 1981, pp. 12.03–12.19

R. A. Cordingley, 'British historical roof-types and their members', *Transactions of the Ancient Monuments Society*, NS, Vol. 9, 1961, pp. 73–120

F. H. Crossley, *Timber Building in England*, 1951

Abbott L. Cummings, *The Framed Houses of Massachusetts Bay, 1625–1725*, 1979

C. R. J. Currie, 'Timber supply and timber building in a Sussex parish', *Vernacular Architecture*, Vol. 14, pp. 52–4

H. Deneux, *L'Evolution des Charpentes du XIe au XVIIIe Siecle, 1927 see* C. A. Hewett, 1969

P. J. Drury, 'A mid-eighteenth century floor at Audley End', *Post-Medieval Archaeology*, Vol. 16, 1982, pp. 125–140

C. P. Dwyer, *Economic Cottage Builder*, Buffalo (USA), 1856

D. Dymond, 'A fifteenth-century building contract from Suffolk', *Vernacular Architecture*, Vol. 9, 1978, pp. 9.10–9.11

M. Exwood, 'Mathematical tiles', *Vernacular Architecture*, Vol. 12, 1981, pp. 12.48–53

J. M. Fletcher and P. S. Spokes, 'The origin and development of crown post roofs', *Medieval Archaeology*, Vol. 8, 1964, pp. 152–183

J. M. Fletcher, 'A list of tree-ring dates for building timber in southern England and Wales', *Vernacular Architecture*, Vol. 11, 1980, pp. 11.32–11.38

J. M. Fletcher, 'Tree-ring dates for building with oak timber', *Vernacular Architecture*, Vol. 12, 1981, pp. 12.38–12.40

Sir Cyril Fox and Lord Raglan, *Monmouthshire Houses*, Vols I, II, III, 1951–1954

E. A. Gee, 'The chronology of crucks', *Transactions of the Ancient Monuments Society*, NS, Vol. 22, 1977, pp. 9–27

S. Giedion, *Space, Time and Architecture*, Cambridge, Mass. USA, 3rd ed., 1956

R. T. Gunther, *The Architecture of Sir Roger Pratt*, 1928

R. Harris, *Discovering Timber-framed Buildings*, 1978

R. Harris, 'Poplar crucks in Worcestershire and Herefordshire', *Vernacular Architecture*, Vol. 5, 1974, pp. 24–25

L. J. Hall, 'Vernacular roof-types in North Avon and South Gloucestershire', *Vernacular Architecture*, Vol. 12, 1981, pp. 12.55–12.60

L. J. Hall, *The Rural Houses of North Avon and South Gloucestershire, 1400–1720*, 1983

H. J. Hansen (ed.), *Architecture in Wood*, 1969 (English edition, 1971)

J. M. Harding, *Four Centuries of Charlwood Houses*, 1976

B. Harrison and B. Hutton, *Vernacular Houses in North Yorkshire and Cleveland*, 1984

W. Harrison, *The Description of England*, 1577, edited G. Edelsen, 1968

G. D. Hay, 'The cruck building at Corrimony, Inverness-shire', *Scottish Studies*, Vol. 17, 1973, pp. 127–135

C. A. Hewett, 'Timber building in Essex', *Transactions of the Ancient Monuments Society*, NS, Vol. 9, 1961, pp. 32–56

C. A. Hewett, 'Structural Carpentry in Medieval Essex', *Medieval Archaeology*, Vol. 6–7, 1962, pp. 240–271

C. A. Hewett, 'The Barns at Cressing Temple', *Journal of the Society of Architectural Historians (USA)*, Vol. XXVI, 1967, pp. 48–70

C. A. Hewett, 'The dating of French timber roofs by Henry Deneux; an English Summary', *Transactions of the Ancient Monuments Society*, NS, Vol. 16, 1969, pp. 89–108

C. A. Hewett, 'The notched lap joint in England', *Vernacular Architecture*, Vol. 4, 1973, pp. 18–21

C. A. Hewett, 'The development of the post-medieval house', *Post-Medieval Archaeology*, Vol. 7, 1973, pp. 60–78

C. A. Hewett, 'Aisled timber halls and related buildings chiefly in Essex', *Transactions of the Ancient Monuments Society*, NS, Vol. 21, 1976, pp. 45–99

C. A. Hewett, 'Scarf jointing during the later thirteenth and fourteenth centuries and a reappraisal of the origin of spurred tenons', *Archaeological Journal*, Vol. 134, 1977, pp. 287–296

C. A. Hewett, *English Historical Carpentry*, 1980

J. Hillam and P. Ryder, 'Tree-ring dating of vernacular buildings from Yorkshire', *Vernacular Architecture*, Vol. 11, 1980, pp. 11.23–11.31

W. Horn, 'On the origins of the medieval bay system', *Journal of the Society of Architectural Historians (USA)*, Vol. XVII, 1958, pp. 2–23

W. Horn, 'The great tithe barn of Cholsey, Berkshire', *Journal of the Society of Architectural Historians (USA)*, Vol. XXII, 1963, pp. 13–23

W. Horn and E. Born, *The Barns of the Abbey of Beaulieu at its Granges of Great Coxwell and Beaulieu*, Los Angeles, USA, 1965

F. E. Howard, 'On the construction of medieval roofs', *Archaeological Journal*, Vol. 71, 1914, pp. 293–352

B. Hutton, 'Some Yorkshire scarf joints', *Vernacular Architecture*, Vol. 12, 1981, pp. 12.30–12.37

C. F. Innocent, *The Development of English Building Construction*, 1916 (re-issued with an introduction by Sir Robert de Z. Hall, 1975)

S. R. Jones and J. T. Smith, 'The houses of Breconshire', Parts I to V, *Brycheiniog*, Vols IX (1963), X (1964), XI (1965), XII (1966–67), XIII (1968–9)

S. R. Jones and J. T. Smith, 'Chamfer stops, a provisional mode of reference', *Vernacular Architecture*, Vol. 2, 1971, pp. 12–15

J. F. Kelly, *Early Domestic Architecture of Connecticut*, 1924 (re-issued, New York, USA, 1963)

H. M. Lacey and V. E. Lacey, *The Timber-framed Buildings of Steyning*, 1974

W. Linnard, 'Sweat and saw-dust: pit-sawing in Wales', *Folk Life*, Vol. 20, 1981–82, pp. 41–55

J. Lister and W. Brown, 'Seventeenth century building contracts', *Yorkshire Archaeological Journal*, Vol. XVI, 1901, pp. 108–113

J. McCann, 'The purpose of rafter holes', *Vernacular Architecture*, Vol. 9, 1970, pp. 26–31

J. McCann, 'Reading the timbers', Parts 1 to 6, *The Period Home*, Vol. 1, No. 6 onwards, 1980

R. W. McDowall, J. T. Smith, C. F. Stell, 'Westminster Abbey, the timber roofs of the Collegiate Church of St. Peter at Westminster', *Archaeologia*, Vol. C, 1966, pp. 155–174

W. B. McKay, *Building Construction*, Vols I, II, III, 1944

D. Martin and B. Mastin, *An Architectural History of Robertsbridge*, 1974

R. T. Mason, *Framed Buildings of the Weald*, 1964

E. Mercer, *English Vernacular Houses*, 1975

G. A. Mitchell, *Building Construction and Drawing*, Part I, 18th ed., 1946, Part II, 14th ed., 1944

G. Moodey, 'The Restoration of Hertford Castle Gatehouse', *Hertfordshire Archaelogy*, Vol. 3, 1973, pp. 100–109

J. Moxon, *Mechanick Exercises*, 3rd ed., 1703 (re-issued 1970)

R. Morgan, 'Dendrochronological dating of a Yorkshire timber building', *Vernacular Architecture*, Vol. 8, 1977, pp. 8.09–8.14

R. Morgan, 'Two tree-ring dated cruck buildings', *Vernacular Architecture*, Vol. 9, 1978, pp. 9.32–9.33

R. Neve, *The City and Country Purchaser*, 2nd ed., 1726

H. B. Newbold, *House and Cottage Construction*, no date but *c.* 1920

J. Newlands, *The Carpenters' and Joiners' Assistant*, no date

P. Nicholson, *An Architectural Dictionary*, 1819

P. Nicholson, *The New and Practical Builder and Workman's Companion*, 1823 and 1838

A. L. Osborne, *Country Life Guide to English Domestic Architecture*, 1967

W. Pain, *The Builder's Companion and Workman's General Assistant*, 1762

E. W. Parkin, 'The Old Court Hall, Lydd', *Archaeologia Cantiana*, Vol. XCVIII, 1982, pp. 107–120

D. Pearce, 'Timber-frame Revival', *Architects Journal*, 6 April 1983, pp. 49–67

W. H. Pearsall, *Mountains and Moorlands*, 2nd. ed. 1968

J. E. C. Peters, *The Development of Farm Buildings in Western Lowland Staffordshire up to 1880*, 1969

J. C. Pitts and others, 'G Warehouse, St. Katherine's Dock, London', *The Structural Engineer*, Vol. 53, No. 10, 1975, p. 418

F. Price, *The British Carpenter*, 1733

O. Rackham, 'Grundle House: on the quantities of timber in certain East Anglian buildings in relation to local supplies', *Vernacular Architecture*, Vol. 3, 1972, pp. 3–8

O. Rackham, *Trees and Woodland in the British Landscape*, 1976

O. Rackham, W. J. Blair, J. T. Munby, 'Roof and Floors of Blackfriars Priory, Gloucester', *Medieval Archaeology*, Vol. 22, 1978, pp. 105–123

F. L. Ragg, 'Helton Flechan, Askham and Sandford of Askham', *Transactions of the Cumberland and Westmorland Antiquarian and Archaeological Society*, NS, Vol. XXI, 1921, pp. 174–233

P. A. Rahtz and others, 'Architectural reconstruction of timber buildings from archaeological evidence', *Vernacular Architecture*, Vol. 13, 1982, pp. 13.39–13.47

H. G. Ramm, R. W. McDowall, E. Mercer, *Shielings and Bastles*, 1970

S. E. Rigold, 'Some major Kentish timber barns', *Archaeologia Cantiana*, Vol. LXXXI, 1967, pp. 1–30

D. L. Roberts, 'The persistence of archaic framing techniques in Kesteven, Lincolnshire', *Vernacular Architecture*, Vol. 5, 1974, pp. 18–20

J. H. Roberts, 'Five medieval barns in Hertfordshire', *Hertfordshire Archaeology*, Vol. 7, 1979

Royal Commission on Historical Monuments (England), *Inventories*

P. F. Ryder, *Timber-framed buildings in South Yorkshire*, 1979

R. A. Salaman, *Dictionary of Tools used in the Woodworking and Allied Trades c. 1700–1970*, 1975

L. F. Salzman, *Building in England down to 1540*, 1952

K. Sandall, 'Aisled halls in England and Wales', *Vernacular Architecture*, Vol. 6, pp. 19–27

J. Smith, *A Specimen of Ancient Carpentry*, 1787

J. T. Smith, 'Medieval aisled halls and their derivatives', *Archaeological Journal*, Vol. CXII, 1956, pp. 76–94

J. T. Smith, 'Medieval Roofs: a Classification', *Archaeological Journal*, Vol. CXV, 1960, pp. 111–149

J. T. Smith, 'Cruck construction: a survey of the problems', *Medieval Archaeology*, Vol. 8, 1964, pp. 119–151

J. T. Smith, 'Timber-framed building in England', *Archaeological Journal*, Vol. CXXII, 1966, pp. 133–158

J. T. Smith, 'The early development of timber buildings: the passing brace and reversed assembly', *Archaeological Journal*, Vol. CXXXI, 1974, pp. 238–263

J. T. Smith, 'Norwegian Stave Churches', *Journal of the British Archaeological Association*, Vol. CXXXI, 1978, pp. 118–125

J. T. Smith and C. F. Stell, 'Baguley Hall: the survival of pre-Conquest building traditions in the fourteenth century', *The Antiquaries Journal*, Vol. 40, 1960, pp. 131–151

P. Smith, 'Welsh Aisle Trusses', *Vernacular Architecture*, Vol. 4, 1973, p. 28

P. Smith, *Houses of the Welsh Countryside*, 1975

T. P. Smith, 'Re-facing with brick tiles', *Vernacular Architecture*, Vol. 10, 1979, pp. 33–36

P. E. Sprague, 'The origin of balloon framing', *Journal of the Society of Architectural Historians (USA)*, Vol. XL, Dec 1981, pp. 311–319

R. Taylor, 'Knee Principal Roofs', *Vernacular Architecture*, Vol. 13, 1982, pp. 34–35

A. Thomas, 'Panel-infill in timber-framed buildings', *S.P.A.B. News*, Vol. 3, April/May, 1982

T. Tredgold, *Elementary Principles of Carpentry*, First ed. 1820, Second ed. 1828, 3rd ed. 1840

B. Tyson, 'Rydal Barns', *Transactions of the Cumberland and Westmorland Antiquarian and Archaeological Society*, New Series, Vols 79 and 80, 1979 and 1980

B. Tyson, 'William Thackeray's rebuilding of Rose Castle Chapel, Cumbria', *Transactions of the Ancient Monuments Society*, NS, Vol. 27, 1982, p. 61

B. Tyson, 'The work of William Thackeray and James Swingler at Flatt Hall (Whitehaven Castle) and other Cumbrian Buildings 1676–1684', *Transactions of the Ancient Monuments Society*, NS, Vol. 28, 1983

T. D. Whitaker, *History of Whalley*, 1901, p. 499

E. Wiliam, *Traditional Farm Buildings in North-East Wales, 1550–1900*. 1982

E. Wiliam, 'A cruck barn at Hendre Wen, Llanrwst, Denbighshire', *Transactions of the Ancient Monuments Society*, NS, Vol. 21, 1976, pp. 23–31

E. H. D. Williams, 'Jointed crucks', *Vernacular Architecture*, Vol. 8, 1977, p. 826

D. Yeomans, 'Medieval Roof Structures', *Architectural Association Quarterly*, Vol. 7, Pt 2, 1975, pp. 44–50

D. Yeomans, 'A preliminary study of "English" roofs in colonial America', *Journal of the Association of Preservation Technology*, Vol. XIII, no. 4, 1981, Toronto

Index